Artificial Intelligence Pushing Boundaries: Is It Worth It?

Simplifying AI From the Basics to the Latest Trends

B.B. Jankoski

Table of Contents

© Copyright 2023 - All rights reserved. 7

Introduction 8

Chapter 1: Unraveling the Artificial Intelligence Landscape 11

    Understanding Artificial Intelligence... 11

        Machine Learning (ML) 12

        Deep Learning (DL).. 12

        Neural Networks... 13

    Types of Artificial Intelligence 13

        Reactive Machines... 14

        Limited Memory Artificial Intelligence 14

        Theory of Mind 14

        Self-Aware Artificial Intelligence.. 15

        Artificial Narrow Intelligence.. 15

        Artificial General Intelligence... 16

        Artificial Superintelligence 16

    The Evolution of Artificial Intelligence 16

        The 1950s 17

        The 1960s 17

        The 1970s–1990s 18

        The 2000s–2010s 18

        The 2020s–Today 19

Chapter 2: AI's Transformative Potential.. 21

    Main Applications of Artificial Intelligence 21

        Healthcare Innovations 21

        Financial Sector Advancements... 22

        Enhanced Personal Education 23

        Entertainment and Media 24

        Transportation and Mobility 25

Cybersecurity 25

Robotics 26

The Influence of Artificial Intelligence on Everyday Life 27

Personal Assistants 27

Online Recommendations 27

Communication and Language Translation 27

Smart Home Technology 28

Customer Service 28

Weighing the Benefits and Drawbacks of Artificial Intelligence 28

Benefits 29

Fast Data Processing and Analyzing 29

Eliminates Human Error 29

Handles Mundane Tasks 29

Able to Handle Risky Tasks to Prevent Sacrificing Human Life 29

24/7 Availability 29

Drawbacks 29

Costly 29

Diminishes Human Creativity 30

Reduction of Jobs 30

Lack of Emotion and Personal Touch 30

Requires Monitoring 30

Ethical Concerns 31

Chapter 3: Top Trends in AI Tools 32

What Started This Artificial Intelligence Hype? 32

Top Artificial Intelligence Trends 33

Generative Artificial Intelligence 33

Explainable Artificial Intelligence 34

Embedded Artificial Intelligence 34

Model Module 34

Data Module 34

Computing Power Module 35

Democratized Artificial Intelligence 35

Multi-Modal Learning.. 35

Computer Vision.. 36

Deepfakes... 36

Artificial Intelligence for Good 37

Choosing the Right AI Tools 37

Productivity 37

Chatbots.. 38

Content Creation 38

Project Management.. 39

Writing... 39

Studying 40

Chapter 4: Ethical Challenges of AI... 42

Ethical Concerns Voiced by Tech Industry Leaders 42

Privacy Violation.. 43

Bias 43

Lack of Transparency.. 43

Lack of Emotional Intelligence 43

Hindrance of Societal Advancement 44

Limit Human Willingness to Make Difficult Choices....................................... 44

Unfair Outcomes 44

Harmful Events Perpetuated by AI's False Predictions 44

Threat to Job Security 45

A Thrill Sport for Criminals 45

Privacy Concerns and Data Ethics in AI 46

Data Collection and Consent 46

Data Security... 46

Data Anonymization 47

Bias and Discrimination... 47

Surveillance and Tracking 47

Data Ownership and Control.. 47

Ethical Data Sourcing...48

Addressing Bias in Artificial Intelligence Algorithms.................................48

In What Way Is AI Biased? 48

How Can Bias Be Addressed?..49

Transparency and Explainability 49

Fairness..49

Diversify the AI Field..49

User Data Rights 49

Proper Guidelines and Procedures..49

Chapter 5: The Role of Humans in the Age of AI 50

Developing and Designing AI Systems...50

Using the Technology to Improve Our Lives...50

Governing the Use of AI 51

Protecting Ourselves and Others From the Risks of Artificial Intelligence 52

Explore Open-Source Artificial Intelligence Platforms With Caution..........53

Be Selective About What You Share Online 53

Do Not Save Your Login Details on Online Platforms 53

Use Strong Passwords..53

Do Not Ignore Potential Bias 54

Ensuring Responsible Use of Artificial Intelligence...................................54

Obtain Informed Consent From Users 54

Freedom to Opt In or Out 55

Debiased and Explainable AI 55

Transparency on Data Privacy and Involvement of Third-Parties 55

Continuous Learning and Training 55

Health and Wellbeing Must Be Prioritized 56

Chapter 6: The Future of AI..57

What Does the Future Hold for Artificial Intelligence?...............................57

The Future of Jobs in the Age of Artificial Intelligence 58

Eliminating Some Jobs..58

Creating New Jobs in the Process 59

Changing the Nature of Work.. 59

The Impact of Artificial Intelligence on Society 59

Changing the Way We Interact With Each Other 60

Challenging Our Beliefs 60

Redefining Our Values... 60

Creating New Social Norms 61

Where Does That Leave You? 61

Integrating Artificial Intelligence in Your Everyday Life........................ 61

Deciding How Much Personal Information to Share............................. 61

Being the Responsible User of Artificial Intelligence 62

Advocating for Ethical Artificial Intelligence Practices 62

Monitor Your Family's Involvement With Artificial Intelligence............. 62

Conclusion 63

Glossary.. 66

References.. 70

Introduction

The current exponential rate of artificial intelligence (AI) evolution is not only alarming but calling for immediate attention if you do not want to wake up to a world you do not even recognize. With experts predicting that the AI global market capitalization will surpass $190 billion by 2030, this indicates a shocking rate of adoption of this highly disruptive technology. It is also expected to make up over $15 trillion of the global economy in the same year (Matleena, 2023). Watching news and witnessing the AI without having a basic understanding of the AI fundamentals can make anyone lose interest because of the perceived complexity of the subject. If this describes how you have been feeling anytime you hear about AI, then you are in the right place because I am about to break down the topic for you.

First things first, let us demystify artificial intelligence so that you can have a quick grasp of the idea and see how far your intrigue will reach. In simple terms, artificial intelligence is the intentional attempt to recreate human intelligence in non-living or man-made machines and computer systems. It means trying to transfer the ability of humans to think, learn, and make decisions to inanimate objects like language models and robots. Language models are computer systems trained on large data sets to use the input data, also known as a "prompt," to produce a human-like responses. These responses can range from predicting the next word or understanding a prompt to produce a desired outcome such as a solution to a problem, language translation, or a text summary. Robots, on the other hand, while they are part of artificial intelligence models, are usually physical machines created for a specific task such as helping in assembling equipment in the industrial setting, cleaning your home, offering service to clients, or exploring territories that are dangerous or risky for a human being.

Before we dive deeper into the topic, allow me to make a bold statement and tell you that you have already used artificial intelligence in one way or another, regardless of whether you were aware or not. If you have asked Siri, Alexa, or Google Assistant for anything, from giving directions to personalized questions or even stating any command, you have definitely used AI. You may have also followed certain social media accounts or watched relatable movies because you followed an online suggestion which was manipulated through AI algorithms to bring you something similar to your previous interests. This proves that despite that the current AI frenzy started recently with the application of generative AI—a type of AI trained to create content, from texts to images, audio, and videos using large datasets to learn and identify patterns as well as understand data relationships—and the trend behind OpenAI's ChatGPT, the technology has a pre-existing history. Although in the past, AI seemed like a complex subject reserved only for highly scientific conversations between technology experts, its current applications in multidimensional scopes indicate that it can be used in more settings than most people anticipated. It is not only revolutionizing industries but able to assist even the average "Joe" to improve their work efficiency and enhance the performance of their services.

Like most people, you may have unknowingly used AI and are only now concerned that the technology is reshaping our world at a more exponential rate than we can keep track of or even control. Sure, most AI applications are highly efficient and effective, as AI helps solve complex problems that would usually take human comprehension and decision making long. Take, for example, its ability to analyze huge amounts of data within a split second, which is something completely baffling. This unmatched processing speed enables companies and individuals to streamline workflow, automate mundane tasks, and eventually, improve performance while saving time and money. However, you may be wondering; at what cost? As the world embraces AI and as the technology continues to push boundaries, what are the consequences? Is it all worth it?

The primary concern regarding this technology is the ethics behind it, as giant companies continue to explicitly use your data without your approval or knowledge. For instance, have you noticed that if you browsed for a certain item online, the next thing you get are adverts tailored around that particular search just to bring your interest even closer? Look at how your social media feed is structured; if you have followed a page on cute babies or scientific content, you start seeing similar pages that you do not even follow. As I mentioned above, all these are the results orchestrated by AI algorithms.

Artificial Intelligence Pushing Boundaries, Is It Worth It? is a comprehensive guide to help you navigate the world of AI. In this power-packed book, you will learn every important detail you must know about this technology, from its narrow inception to its current multidimensional application. The book will further delve into the core benefits of using AI in our everyday lives, the potential risks, and precautions we must responsibly adhere to. I will also include the top trending AI tools and how you can incorporate them to your systems or personal application. Not only will you learn how to responsibly apply this technology, you will also know how to tackle the ethical concerns and play a better role in educating those around you to be just as responsible. With the arsenal of tools shared within this book, you will be able to choose your own level of involvement with AI and make informed decisions rather than simply choosing oblivion and suffering the consequences, while there was something you could have done.

If you have previously been sitting on the sidelines due to the belief that this technology is complicated or that you want nothing to do with it, I invite you to dive into this simplified guide on AI so that you can, at least, gain basic understanding of what it entails. As I mentioned, you may have unknowingly interacted with AI; how about you get informed so that the next time you interact with the technology, you know the impact of your actions? The answer to this question lies inside this book; see you inside!

Chapter 1: Unraveling the Artificial Intelligence Landscape

Before diving into the main question of adopting or rejecting artificial intelligence, it is crucial that we level the playing field and reel you in with the basics of this technology. I will start with demystifying the core terminologies in simple terms while associating their application on everyday life. With a basic understanding of AI fundamentals, it will be easy for you to connect the dots as you encounter the technology's application. Hopefully, shedding some light on the topic will give you enough to decide on your level of interaction with the technology. This chapter is your north star regarding the history of AI, from its inception to its future .

Understanding Artificial Intelligence

A more technical definition of artificial intelligence (AI) would be that it is a simulation of human intelligence by computer systems and machines (Burns, 2022). A simpler version of AI means giving computers and machines the ability to think and learn in a way akin to how humans do. It involves creating programs and systems that can understand information, make decisions, and solve problems, all without being explicitly programmed for each specific task. Think of AI as a smart assistant: Just like you might ask a friend for a recommendation or advice, you can ask AI systems different questions—or give them tasks to do—and they will use their knowledge and learning ability to help you find answers or complete given tasks. They can perform various activities like understanding your voice commands, recognizing objects in images, playing chess, suggesting movies you might like, or even helping doctors diagnose diseases.

The goal of AI is to get the computer systems, machines, or AI models to perform activities humans perform, only better, to improve efficiency in areas like speech recognition, image processing, language learning, decision making, and problem solving. This ability is achievable due to the ability of AI models to learn, even if it means being programmed to know certain information to be able to use it later. Another simple way to understand AI is to think of it in human growth stages. From how a baby learns to identify objects up to when they recall their names, and even learn distinguishing features, until they get so smart that they can even start teaching you about relative objects, AI enables machines to develop knowledge and get smarter as they interact with more information. Within the AI concept lie its subsets, known as machine learning, deep learning, and neural networks, all which I break down below.

Machine Learning (ML)

As the name suggests, machine learning is the ability of a machine or computer system to continually learn from data provided and be able to refine its suggestions rather than simply guess based on the core programming it has memorized. It simply means the capability of the machine to use new data to get smarter as opposed to remaining with the same level of knowledge by sticking to only what it was taught during its training. This is achieved by coding the machine to learn, understand, improve, and later use new data to identify objects it has been taught and beyond. For instance, if you have taught the machine that a bird is a two-legged organism with a head and a long beak by showing it numerous pictures with birds, it can then understand this information and be able to use it to recognize and distinguish a bird from other animals like cats and dogs. As the machine gets exposed to more data, it also learns that there are different types of birds with different beak sizes. The more it interacts with new data, it gets smarter to a point where it can tell different birds apart beyond the core structure of a bird simply having a head, a beak, two legs, and wings.

Machine learning is an integral component of AI that enables optimization within organizations. It involves using technical statistical techniques to enable systems to analyze and interpret large amounts of data, identify patterns, and make informed decisions accordingly.

Deep Learning (DL)

Called "deep" because it is multi-layered just like the depth of our brains, deep learning is a branch of machine learning that uses neural networks to process and analyze complex data built on simple understanding. Deep learning algorithms are designed to automatically learn and extract meaningful patterns from huge amounts of data. While they are trained to perform cognitive tasks such as recognizing images, identifying patterns, distinguishing voices, and processing natural language, deep learning AI algorithms can lead to new data without fully revealing how it came up with such information. Similar to how you can come up with an idea without knowing what triggered it, some information from deep learning can be fascinating but unreliable due to its lack of basis. Can we, as humans, entirely trust all the data and predictions provided through deep learning? That is one of the main concerns which we will navigate through later. At its core, deep learning AI plays a vital role in applications like helping self-driving cars smoothly navigate the road by detecting red lights, other cars, and crossing pedestrians. It also enables virtual assistants like Siri and Alexa to understand human commands through speech recognition. Another integral application of deep learning is in the health care sector where algorithms help doctors in spotting diseases in medical scans like X-rays and MRIs.

Neural Networks

Inspired by the structure and function of our human brains, neural networks are a type of AI model consisting of interconnected nodes, also known as "neurons," or tiny cells responsible for processing and transmitting information. The most common application of neural networks can be seen in natural language processing, image, speech, and pattern recognition. As a branch of deep learning, neural networks are designed to learn and make predictions based on input data, solve complex problems, and improve their performance over time by adjusting their internal settings called "weights" and "biases." When these AI models make a wrong prediction or give you incorrect output, you correct it, and it takes that information to improve its accuracy in the future.

Types of Artificial Intelligence

Depending on different capabilities, you will now learn the various types of AI and how it is meant to continue evolving. From weak AI, which is heavily reliant on humans, to strong AI systems that are not only self-aware but possess the intelligence that surpasses human understanding, the tech space has come a long way. These types of AI indicate a transformative technological journey from the past to the future.

Reactive Machines

This is the beginning of AI, where machines are programmed and trained to perform specific tasks. Reactive machines have no memory and, thus, cannot use past experiences to improve their performance. They tackle a given challenge anew every time. Unlike with the subsequent AI types that are able to learn from data to improve their efficiency without any reprogramming, the reactive machines' code would need to be adjusted for them to perform new tasks or behave differently from the way they were trained. For example, you need to program your washing machine for every new laundry load because it has no memory of its previous task. Although reactive machines come with limitations, they created a foundation for other types to be improved upon.

Limited Memory Artificial Intelligence

As hinted in the name, limited memory AI is able to store limited amounts of data within a short period. Unlike reactive machines, limited memory AI can use past experience and new data to make predictions and come up with decisions using their built-in memory. For instance, a self-driving car knows that it has to stop at a red light and can sense other cars and pedestrians to make decisions as traffic changes. It can use this data to analyze oncoming traffic to determine whether to change lanes or adjust its speed. While this is an improvement from reactive machines, limited memory AI is also much limited compared to its successors, as it cannot independently improve itself without human intervention.

Theory of Mind

Theory of mind AI is the advanced stage of the technology where machines are able to read and analyze human emotions and respond accordingly. Rather than simply carrying out instructions, theory of mind AI machines can sense your tone, mood, or even intentions and help you navigate through the given tasks with emotional reaction as well. For instance, imagine if Siri or Alexa could read your emotions when you are frustrated as you give a command, then, instead of these virtual assistants giving a generic and non-personal response, they give output that considers your frustrations. While this type of AI is still in its conception, or near-infancy phase, it is highly anticipated, as we can then have machines that get us and collaborate with us in a harmonious way. Having theory of mind AI can also be helpful to the neurodivergent community, as it can be programmed to read facial expressions and mental states to ease any social communication challenges.

Self-Aware Artificial Intelligence

Also referred to as AI reaching a point of singularity, self-aware AI means exactly what the name implies: Not only are the machines in this category able to sense the feelings and emotions of others, but they are also aware of themselves. Nobody knows all the details regarding this final developmental AI stage with regards to when it will be or even if it will really come to this point. The only consolation is that it is currently only a futuristic concept with no proof that AI will reach this level. However, you will agree with me that it sounds pretty scary having machines that are able to react emotionally beyond their programming. It means machines are likely to deviate from given instructions and decide, on the spot, to disobey humans as per their original design. Who knows what they will do if they feel violated? Will they retaliate? For now, let me leave your mind to marvel at the infinite possibilities.

Artificial Narrow Intelligence

As the name suggests, artificial narrow intelligence (ANI) machines are designed to perform a select number of tasks limited to a narrow capability. They cannot perform wide ranges of tasks beyond the ones they are specifically designed for. ANI machines are also known as weak AI because they cannot independently learn, think, or improve themselves beyond their original design. Like other earlier versions of AI such as reactive machines or limited memory AI, ANI machines are trained to only excel in a particular cognitive ability. If you have used facial recognition biometrics to unlock your smart phone or voice commands to give a virtual assistant like Siri or Alexa instructions, you have used this type of AI. Self-driving cars also fall in this category. As you can tell by how long these applications have been in use, this technology has been around longer than the current trend, yet most people never bothered to question it.

Artificial General Intelligence

Also known as "human-level" AI, artificial general intelligence (AGI) is the next level of AI capability where machines cannot only be programmed to do specific tasks but can do a wide range of functions, learn, think, and perform at the intelligence level of humans. It is considered as strong AI, as it can understand and adapt to any task a human is capable of doing.

Artificial Superintelligence

Although it is currently inclined and limited to mere theory, artificial superintelligence (ASI) is designed to be a further integration of self-aware AI; AI that not only understands and thinks independently like humans but surpasses human intelligence. If you look at the way it is fictionalized in science fiction and action movies like *I, Robot* and *Ex Machina*, to mention a few, superintelligence AI is more likely to stage a takeover. Besides the sci-fi aspect, nobody knows if this will ever be achievable. However, the exponential rate at which the technology is evolving calls for concern, especially for the upcoming generations. What world will they come into? Some experts even predict that ASI is nearer than we anticipate. Considering the current developments toward making AGI a success, their theories are not baseless.

The Evolution of Artificial Intelligence

From almost a century ago, artificial intelligence has been making headlines for its perceived capabilities to simplify and improve our lives. Although it came with its own complexities, the ever-changing technology has fascinated scientists and futurists to think outside of the box. Until now, smart thinkers are in the business of inventing and enhancing machines that would, one day, behave like humans and perform tasks that we know are only limited to us. This section paints a vivid picture of how far we have come with this technology to date.

The 1950s

Despite only making major strides since the inception and popularity of generative AI, the concept of simulated human intelligence dates as far back as the 1950s. If we look beyond the narrow meaning of AI and focus on its components like language processing and making predictions, we can say that the ability to argue a theory can be taken back to the ancient mathematicians who used complex math language and formulae to prove their points. The invention of computers as human assistants doing complex tasks also counts as the groundwork leading to the epic artificial intelligence evolution as machines got smarter and more efficient.

Although the first work representing what we know today as AI was introduced in a publication by Alan Turing during the Turing Test—a test conducted to check if a computer can have a convincing human-like conversation in such a way that you cannot tell if it was a computer or human being—in 1950 (Karijan, 2023), the term "artificial intelligence" was only coined in 1956 by a renowned American computer scientist, John McCarthy, during a Dartmouth College conference. Attended and supported by other pioneers like Marvin Minsky, Nathaniel Rochester, and Claude Shannon, who had already started the foundational work toward the advancement of the technology, the conference marked the birth of AI as a field of study (Dartmouth, 2006).

The 1960s

The 1960s era saw the growth in the first AI programs which could do simple things like playing games and solving math problems. With developments in neural networks, machine learning, and programming languages created by legends leading the innovation, AI gained incredible popularity among developers. Early fascination led to successful AI systems designed to beat humans in mind challenging games such as chess and checkers. Other early AI applications included language translation.

In 1966, another ground-breaking milestone was reached when Joseph Weizenbaum created Eliza, the first and the most celebrated computer program that paved the way for today's chatbots. With the capabilities of holding an engaging, human-like conversation, Eliza increased developers' fascination with the technology, and they believed in the possibility of AI emulating human-like emotions. In the same year, the world's first intelligent robot, Shakey, was developed by the Stanford Research Institute. As the core inception behind self-driving cars and drones, Shakey brought a fusion of AI with computer vision, navigation, and natural language processing (Karijan, 2023).

The 1970s–1990s

After Marvin Minsky and Seymour Papert released *Perceptrons*, a book that highlighted the limitations of neural networks in 1969, there was a decline in the field's research. This was exacerbated by the report published by James Lighthill in 1973. Lighthill's report, which revealed the general survey on AI, led to a significant withdrawal from the British government toward supporting AI research. All these negative sentiments were the start of the longest technology nap, also known as the "AI winter." By the start of the 1980s, all the hype and interest in AI had died, as companies that were pioneering machine developments were also experiencing decline. The term "AI winter" is attributed to Marvin Minsky and Roger Schank, who warned businesses in 1984 that AI hype was dying, and that the industry was fast approaching its collapse. Indeed, three years later, the walls of the AI structure started crumbling down.

Although the AI winter came with a lot of distrust in the system when investors started losing money and reports indicated the limitations of neural networks, things shifted in 1989 when Yann LeCun, Yoshua Bengio, and Patrick Haffner demonstrated a positive side of neural networks. In particular, the trio revealed the possibility of using convolutional neural networks (CNNs) to recognize handwritten characters and solve real-world problems. The beginning of the 1990s came with innovation around machine learning and advancement in robotics. Another positive event that worked in the technology's favor was the emergence of the World Wide Web, which provided a large data supply to AI systems. This era ended on a high note when IBM's "Deep Blue" defeated the world's renowned chess champion, Garry Kasparov in a chess rematch in 1997. It caused further research and exploration around AI's capabilities.

The 2000s–2010s

The beginning of the 21st century indeed reflected that not only was AI winter a thing of the past, but advancements came at full speed as computers became super fast and data was everywhere. More and more robotic companies and individuals were testing out the artificial mobility and intelligence of their new creations. This era also saw robots designed to navigate space and other planets and robots that could interact with humans and read emotions. One iconic AI moment in this field can be referenced to February 2016, when a Hong Kong-based company, Hanson Robotics, released the first humanoid robot, named "Sophia." While there are arguably several attempts to create humanoids, Sophia is the first social robot created in the shape of a woman. She has also been given an identity, an honorary citizenship in Saudi Arabia, and other privileges such being elected as the innovation ambassador for the United Nations Development Programme (UN News, 2023). As a social robot, Sophia has the ability to engage in human-like conversations, remember interactions with people, display various facial expressions and gestures, crack jokes, and even possess artistic features like singing and drawing.

Moreover, this era also came with enormous advancements in social media and online presence. With the majority of the public now feeding social media platforms with free data, things became significantly easier for AI to advance in making predictions regarding our preferences. I am certain that you have wondered how your search history pops up even when you are on a different platform. For instance, if you indicated your interest to buy a property in a certain area, you will notice that you start seeing advertisements or related group information either on Facebook, YouTube, or any other platform. I was recently searching for medical aid quotes, and lo and behold, I cannot scroll my feed for a minute without quotes from different providers popping up. That is what AI algorithms are doing; bringing my search nearer and possibly influencing my decision. With social media platforms selling our data to third parties, the issue of data privacy became a major concern that we are currently facing.

The 2020s–Today

The current AI wave can be attributed to OpenAI's brainchild, ChatGPT—a generative pre-trained transformer chatbot that has been provided with a massive pile of data and trained to process human-generated text (prompt) and respond with human-like output. ChatGPT became freely accessible to everyone since November 2022, and over one million users signed up within five days, making it one of the most adopted technologies of all time (Buchholz, 2023). By early 2023, ChatGPT had over 100 million cumulative users, and with further updates and advancements on the capabilities of generative AI, even those who have been slumbering all these years when the technology progressed started paying attention. About 35% of companies are now using AI, while 42% mentioned that they are exploring the possibility of integrating the technology in their systems (Zauderer, 2023).

Another AI-powered social media platform, Threads, developed by Meta in 2023, broke ChatGPT's record when it reached the one million sign-ups milestone in less than one hour. Threads is also one of the first social media platforms to put new AI-powered features like translations and long conversation summaries to the test. As you will see with various AI applications detailed in the next chapter, this technology is becoming mainstream. The ability to save time, improve performance, assist in medical procedures, and recent discussions about regulations and ethical advocacy, the AI landscape is burgeoning at an alarming rate. Whether the masses fully understand their role in AI's integration is another story. This is my most pertinent reason to bring it to your attention so that you understand and choose your level of involvement with the technology, knowing its implications.

Chapter 2: AI's Transformative Potential

Since its inception until now, AI has drastically changed the way we look at and go through challenges. It has shifted our perspective on things that we thought were humanly impossible. It is difficult to ignore a technology that is dedicated to improving how humans work and play. However, as much as it has proven itself as a formidable force, AI still requires responsible operators. This chapter sheds more light on the groundbreaking applications of the technology. We will also focus on its influence on our daily lives. As we know that every good thing has the potential to have a weak point, we will close this chapter by exploring some of the common benefits and drawbacks of AI.

Main Applications of Artificial Intelligence

From saving lives through expert medical record analysis and predictions to undertaking risky interplanetary missions, AI applications have changed the way we do things and will continue to amaze us. From personal assistants that anticipate our needs to algorithms that curate our personalized content, the technology continues to revolutionize the way we work, communicate, create, and interact with our world.

Healthcare Innovations

With a proven special ability to conduct detail-oriented data processing and analysis of complex medical records like X-rays, MRI scans, and other tests, there are controversies that AI is outsmarting human doctors. After several accurate disease diagnoses, early cancer detection tests, and suggestive patient treatments that human doctors missed, some medical practitioners voice their fear of AI possibly leaving them out of jobs (Satariano & Metz, 2023). Arguably, AI's ability to save lives is one of the most groundbreaking technological advancements in the world. Well, maybe we are still a bit further away from AI completely replacing human intelligence, as their incredible findings still need a qualified eye to prove authenticity and accuracy, particularly in regard to sensitive data like medical reports.

From the University of California San Francisco, Dr. Maria Wei argued that AI systems still need intensive training using various racial groups so that AI models have huge datasets in different conditions to be able to accurately detect melanoma. She mentioned that this will help avoid unnecessary biopsies while still training the machines to improve in analyzing every minute detail (Melanoma Research Alliance, 2022). This life-saving AI application is one of those that makes it hard to completely ignore or disregard the technology. While AI's role in healthcare innovations is still in its infancy, it is a convincing reason to further explore more potential applications of this technology for the good of humankind.

Financial Sector Advancements

Similar to medical data in terms of sensitivity, finances are an avenue that require handling with utmost trust, care, and confidentiality. The financial industry is at the most risk of fraudulent activities and misuse of people's personal details. AI plays a crucial role in helping this sector with fraud detection, identity verification, and risk assessment. For instance, insurance companies are able to compensate clients accurately without overpaying or underpaying for the reported risks and costs of damage. Additionally, using AI to verify borrowers' credit scores and financial position ensures that lenders offer loans with less default risks due to borrowers being unable to afford loan repayments.

Banks are also able to use AI to detect any suspicious behavior and effectively combat fraud. From monitoring suspicious fraudsters' spending habits to identifying any clues to link them to possible financial crimes, AI systems reduce the investigative workload so that financial houses can focus on quality service delivery and ensure the security of their clients' funds. Improving customer self-service through chatbots has also made banking so much easier. With personalized communication and client-specific chats, there is little to no need for you to run to the bank and queue for service. This enables you to conveniently bank from anywhere. It also allows banks to focus on improving other services, knowing that most of the admin work and mundane tasks like identity verification or know your customer (KYC) compliance has been entrusted to bots.

Moreover, AI trading bots have grown in popularity due to their efficient analysis of financial markets and *mostly accurate* predictions. High-frequency trading relies on the ability of AI trading bots to analyze economic news (fundamental analysis) as well as trading charts (technical analysis) within a short period of time and be able to predict market direction. Market analysis and trading decisions usually take humans longer due to our inability to access relevant information and verify trading parameters before executing trades. Human beings are also emotional traders in most cases, while trading bots use algorithms and quantitative analyses to make fact-based decisions rather than those influenced by feelings. While trading bots have contributed to profitable investment choices for technology savvy investors, it is worth noting that AI application in the financial sector still carries an inherent amount of risk: It still requires responsible monitoring instead of fully trusting the bots to offer the best services or financial advice.

Enhanced Personal Education

Compared to the obsolete teaching techniques where learners are offered generic exercises, AI-powered education can be personalized to help individual learners from their specific learning capabilities. No two students are alike and expecting them to comprehend the same information at a similar speed and understanding level, as is common with traditional education, is setting slow learners and those with special needs up for failure. Luckily, AI is able to identify individual learner's comprehension level, note their struggles, and customize a different approach to meet them at their point of need.

Using AI-powered education techniques also enables educators to enjoy their work knowing that they are reaching out to every learner according to their comprehension level. This brings less frustration and worry to them and allows them to come up with fun-filled effective lessons and focus on character building rather than having a job that truly feels like a chore. Additionally, rather than having teachers doing mundane tasks like grading students, AI can automate such activities and provide feedback that educators can later use to help individual students. Educators also have more time to connect with students and positively influence them to pursue their dreams and conquer their fears.

The main concerns with AI-powered education is affordability and privacy violation. Starting with the costs, not every school or learner can afford to have AI structures integrated into their existing learning processes. This means that on a large scale, students with access to AI-powered learning will be at a higher advantage than those from underprivileged backgrounds. This injustice becomes most visible when these learners enter the corporate world and there are evident disparities that lead to opportunities being handed to advantaged individuals. The issue of privacy also has to be addressed because of the possibility of schools misusing student personal information to track their out-of-class activities and conduct. While this information can help give the educators the reality of the learners' home situation, it can also embarrass learners if educators call them out based on their backgrounds.

Entertainment and Media

There is a significant rise in the application of generative AI in the entertainment industry. An amazing feature can be seen with actors having a digital twin by training AI models with data from the previous work actors performed. AI models are then able to mimic the actor's voice, gait, gestures, and emotional reactions by recreating certain scenes or new ones altogether. Digital twins can further be useful to aging actors who may want to continue their screen work but are physically unable to. For instance, legendary actors like Bruce Willis, Gene Hackman, Michael J. Fox, and Christina Applegate are some of those who recently had to step down from the careers they love due to physical and mental ailments. Imagine that AI has the capability to give actors an extended screentime and immortality. However, the major concerns are ownership and copyright issues. Who exactly owns the digital twin? Who protects its rights? And what would be the role of the actor going forward? These are some of the questions to consider to avoid some actors possibly being forced out of the game should their digital twins show more potential.

Moreover, generative AI is also widely used in the music industry with artist voice cloning and the ability to feature superstars in new songs. As with the film industry's copyright issues, the application of this technology has only been exercised without prior permission of the famous singers in recording AI-generated music. The same applies with regenerated or plagiarized art. The AI application in entertainment and media requires serious policing to ensure that artists' hard work is not being pirated or used without compensation. There are various ways in which the artists can use the technology for their own benefit to streamline their workflows and focus on adding a personal touch while leaving AI to do the mundane tasks.

Another entertainment area where AI application has been brought to life is gaming. With enhanced communication and collaboration with in-game characters, game players are able to explore virtual worlds using augmented reality and virtual reality. This makes playing video games feel real, giving players an authentic and exhilarating experience. As a media consumer, you may have already noticed AI-powered media and entertainment suggestions are customized according to your preferences. Like having a best friend who pays attention to what you like and delivers it, AI algorithms are able to monitor your movie, music, and content choices and bring similar content closer to your search.

Transportation and Mobility

Already adopted by behemoth logistics and product distribution companies like DHL, FedEx, and Amazon, there is a burgeoning AI application in mobility and transportation. Delivering products from the manufacturing sites to the end-users is an integral part of any business. If there are any delays, damages, or inconveniences, the business not only loses money to remedy the situation but also loses the trust and loyalty of their clients. The manual operation of a logistics company is an obsolete way of doing things. Therefore, this is an area where deploying the technology continues to marvel. The ability to track delivery transport and monitor quality assurance at loading and offloading stations minimizes errors and improves service delivery. Using computer vision, delivery vehicles are able to detect traffic and optimize routes to save time and reach their destinations faster.

Cybersecurity

Currently at around $10 billon, the market capitalization of AI in cybersecurity is estimated to be above $80 billion by 2030 (Sajid, 2023), indicating the rise in demand and implementation of the technology. With a growing digital culture where most of our daily operations are conducted online, online data theft has become a goldmine for cybercriminals and bad actors. People are being bullied online and personal information and identities are stolen and misused at an alarming rate because cybercriminals are advancing faster than online law enforcement. AI is the nearest and best solution to curb these problems because of the ability for early malware and phishing detection, breach risk detection, and task automation. Instead of waiting for a threat and only dealing with it later, as is common with manual cybersecurity, AI models are trained to sense an attack before it actually occurs and put up firewall or network defense mechanisms in place.

Moreover, director of cybersecurity advocacy at Deep Instinct, Chuck Everette, boldly claimed that AI-powered malware detection systems are about 80–92% efficient while signature-based ones only have the efficiency rate of 30–60% in preventing cyber threats (Sajid, 2023). Using computer vision and neural networks has proven to be an effective exercise in cybersecurity. Machine learning-based security systems also ensure that threats and security breaches are dealt with even before they occur due their ability to process and consolidate large amounts of data that would be tedious to humans. Manual detection of threats would take lots of time and resources, which would defeat the purpose looking at the rate of new vulnerabilities. By the time humans deal with an old problem, how many new threats would have penetrated the system? With the ability to automate threat detection and employ necessary counterattacks, this is another area where AI deserves a thumbs up and serious consideration. However, it is worth mentioning that using AI in cybersecurity is not all rosy; cyber attackers can still obtain access to AI models' training material and use it to manipulate AI systems. Another downside is that AI models are fed with real-time data that is often sensitive, further raising concerns for privacy issues.

Robotics

With implementations dating as far back as the 1960s, the field of robotics is arguably one of the first applications for AI. It has simply grown into different industries. Although we have seen a rise of robots in fiction films, this is one area where AI application can be seen in a multidimensional scope. Robots can be assigned to any task in any industry. Our main focus on this section will be on physical robots that have mobility access. It is estimated that about 400,000 robots are introduced yearly, with about three million currently powering the industries in the world (Sajid, n.d.).

From assembling vehicles, lifting heavy machinery, and performing all tasks that would require more human effort than necessarily possible, AI in robotics has made major strides on the industry level. However, that is not all where it is useful. You may have also seen a rising trend in service robots. The COVID-19 era populated contactless service as companies employed service robots to take people's food orders in restaurants, deliver food to respective tables, greet customers, harvest crops, weld machinery, and efficiently perform most automatable duties. These robots already existed prior to the pandemic even though they were deployed only on a small scale or by early technology adopters, irrespective of whether they were tech or non-tech sectors.

Besides efficient productivity and better customer service, another fascinating area where robotics shine is in undertaking risky tasks like exploring territories beyond our current ability. Curious people and scientists have tried exploring other planets, like Mars, that they cannot personally reach with the current technology. And yet, they have been able to successfully send robots on a mission to explore and take samples in search of signs of possible life on the red planet. The arduous journey to explore any possibilities of interplanetary life can only be embarked on by machines, and since Mars Rovers made a successful landing, scientists are able to further study Mars and other fascinating beyond Earth missions.

The Influence of Artificial Intelligence on Everyday Life

AI has become a part of our daily lives, simplifying tasks, enhancing communication, and making our homes smarter and more convenient. It is like having a helpful friend or personal aide who is always there when you need assistance or a recommendation. So, the next time you ask your AI assistant for the weather or shop online, remember the magic behind the scenes.

Personal Assistants

With the ability to set your reminders, answer questions, make hands-free calls, and even tell you jokes, AI-powered personal assistants are like your digital companions. Imagine asking, "Hey, Siri, what's the weather today?" and getting an instant response, helping you plan your day. You can ask your personal assistants all sorts of commands, and they will respond. Even when they cannot provide a concrete answer to your prompt, they can tell you their limitations and suggest resources where you may get better help.

Online Recommendations

When you shop online, AI algorithms quietly work in the background, analyzing your preferences based on your past choices. They then suggest products or content that you are likely to love. Think of it as having a friendly shopping assistant who knows your taste and shows you the perfect pair of shoes you did not even know you wanted. Just bear in mind that this might increase impulse buying; however, it is a useful tool to a disciplined shopper who wants to buy some items without having to know all product names. Even if you wanted a certain range of items and did not know that there were more items to make the set complete, AI algorithms will bring it to your fingertips.

Communication and Language Translation

AI helps you communicate with people from all over the world, even if you do not speak the same language. Chatbots and translation tools can instantly convert your words into another language. Imagine chatting with a friend from Japan without knowing Japanese. AI can translate your messages so you can have a meaningful conversation and connect better with others.

Smart Home Technology

With AI-powered smart home devices, your home becomes like a wizard's castle. You can control your lights, thermostat, and security systems with your voice or smartphone. With home integration, your home obeys your commands to switch on the lights and regulate temperatures to your liking.

Customer Service

Whether you or your customers have a question or need help with a product, AI chatbots and virtual assistants are there to assist. They can answer common questions and guide you through troubleshooting. They excel as customer service agents, available 24/7, to help track a package or find answers to frequently asked questions. This highly responsive feature makes any online platform efficient for both customers and organizations.

Weighing the Benefits and Drawbacks of Artificial Intelligence

Like any innovation, AI has its own areas where it marvels and possibly saves us a lot of trouble and effort. It also has limitations that can harm or diminish society's quality of life. In this section, I lay out both the good and bad of this technology and leave you to weigh your options.

Benefits

Fast Data Processing and Analyzing

Due to their capability to process enormous amounts of data within seconds, AI models are able to read through and analyze any given data in search for clues and answers. This assists in time-sensitive decision making where humans may be hesitant. Where humans are overanalytical or emotional, AI can be result-oriented and swift.

Eliminates Human Error

Trained to fully focus on minute details, AI systems are precise and consistent. This reduces the likelihood of errors that humans might make due to fatigue or distraction.

Handles Mundane Tasks

AI excels at repetitive and monotonous tasks, freeing humans to focus on highly creative and strategic roles where a personal touch is required.

Able to Handle Risky Tasks to Prevent Sacrificing Human Life

As mentioned earlier in the robotics section, AI agents can carry out the most life-threatening missions. Whether it is diffusing bombs or exploring dangerous territories where human lives would be at risk, they accept the call and take one for the team without any hesitation.

24/7 Availability

AI-driven chatbots enhance customer service and provide immediate responses to client queries. They work around the clock without needing breaks, ensuring continuous operation and availability.

Drawbacks

Costly

Although there is currently a rise in free AI tools, you still need to subscribe to premium access to get the best experience and ability to own your created content. For some tools, the cost of entry is not affordable to the average "Joe." As I mentioned in the education section, some colleges and individuals cannot afford to integrate AI tools in their current academic systems. Similarly, AI that helps doctors detect and diagnose hidden medical conditions also does not come cheap. Depending on the size of a company and the complexity level of an AI-powered solution, the cost of AI usually includes software creation, manpower, training, and maintenance. Software costs can range from $30,000–45,000, manpower $25–49/hour, while training and maintenance would roughly cost $8,999–14,999. It can cost up to $100,000 for a mid-sized company integrating a moderate to high complex AI solution tool, with smaller and larger companies seeking low to high complexity levels ranging on either side (RisingMax, 2023). These mindboggling figures indicate that even with highly sought-after AI solutions, the costs might be a serious hindrance.

Diminishes Human Creativity

The extensive use of AI tools increases humans' reliability on machines to create and perform tasks that they used to enjoy and excel in.

Reduction of Jobs

As much as AI applications also come with new job positions, there is a high risk of people being sacked from areas that are easily automated. People who have not yet embraced the technology do not know how to upskill themselves. If automating their current tasks is more efficient and cost effective to the organizations they work for, their obsolete skills are no longer required, and that leaves them out of jobs, thus contributing to high unemployment rates.

Lack of Emotion and Personal Touch

For fields like counselling and caregiving, AI applications would sound generic and impersonal. Some areas require humans to be empathetic and understanding, while AI agents are trained to focus without deviation. Moreover, although some people may argue that generative AI has a human element in it, there are simply things that no robot can personalize. Text generated through GPT and other models may sound like it was written by humans if the prompt instructed the bot to write in an engaging tone. However, without any editing, that text would lack a personal touch in terms of true examples that show empathy and other emotions. Even though the recipient of AI-generated material may be emotionally moved without fully comprehending that it is not generated by a human, the sender knows that it does not make the material truly sentimental. Think of receiving an AI-generated love poem and being moved by the words that are not from someone's deep emotional dwelling but from an emotionless bot that studied millions of poems and regenerated them.

Requires Monitoring

AI systems continually require monitoring and human supervision to ensure smooth and ethical operation. This also adds on to the cost of this technology.

Ethical Concerns

The use of AI raises ethical issues, including bias in algorithms, data privacy, and AI's impact on society, requiring careful consideration and regulation. This is discussed in further detail in Chapter 4.

Chapter 3: Top Trends in AI Tools

Arguably, the technological space is one of the fastest evolving innovations. The top trends I cover in this chapter are not timeless per se because even after publishing this book, I believe some of the trends here might be obsolete. This means you have to be on the lookout and remain engaged, as you will learn more if you are interested in this area. While it has been trending for a long time, the AI field was not everyone's interest, as most people barely understood its developments or purpose, which is unfortunate because we have numerous pressing issues. Even if we were told that AI's advancements would play crucial roles in solving health issues, disasters, and overall improve human efficiency, all these sounded too technical for an average mind to comprehend. However, the current AI wave hooked the masses' attention, as we started seeing possibilities of applying the technology without possessing prior scientific experience or other complicated skills. This chapter paves the way on how we can use the current and upcoming AI tools in our different spheres.

What Started This Artificial Intelligence Hype?

AI has always been making waves since its inception, although some waves did not last long enough to inspire mass adoption. Mostly popular in fiction novels and feature films, the advancement of this technological space has caused controversies. While the current hype can be attributed to various OpenAI creations like DALL-E and the GPT series that have the capabilities of generating human-like output from various prompts, robotics is where most advancements can be referenced. Do you remember Sophia, the humanoid robot and the brainchild of Hanson Robotics? As I mentioned in Chapter 1, Sophia is the first AI robot to be granted honorary citizenship and a chance to feature in transformative conferences and deliver speeches. Having had the opportunity to travel numerous countries and being a guest at renowned TV shows and large platforms, Sophia opened the eyes of many toward the advancement of AI and its capabilities. While she was not the first humanoid, her popularity and wittiness made her one of the people's favorite robots. To other people, Sophia's intelligence, subtle jokes, and controversial sarcasm heightened concerns and fear of the perceived robot-humanity takeover.

In 2023, another humanoid robot, Ameca, who possesses both artificial intelligence and artificial body, added to the current AI frenzy. Ameca is said to be the world's most advanced humanoid, as it is capable of being integrated with the advancing natural language processing, computer vision, and neural networks, among other AI developments. So, while most people only started paying attention to AI now with generative AI, they are only riding on a preexisting wave that started a couple of years ago. The main thing we can be grateful of regarding the latest AI wave is that it demonstrated that we do not need to possess any prior knowledge to understand this field of study that was once regarded complex and reserved for science and technology nerds. As more people adopt this technology and integrate it in their day-to-day activities, the hype continues, and it does not show any signs of dying down any time soon.

Top Artificial Intelligence Trends

Besides the trending specific or individual AI tools discussed toward the end of this chapter, we can look at the disruptive AI trends by categorizing them according to their capabilities, applications, and popularity. The following trends are areas where AI has witnessed the most growth in general.

Generative Artificial Intelligence

Responsible for the current wave of AI, generative AI is designed to help us automate mundane and boring tasks and generate human-relatable content. Trained with large amounts of data, this game-changing type of AI is able to create mind-blowing content from any type of prompts given. It can generate detailed custom images, text, audio, and video output from simple commands and refine them as you also refine your prompts. Generative AI is the most widely used and adopted AI model today, that anyone can use without any extensive background in prompt engineering because it uses natural language. You can collaborate with this technology to produce product descriptions, write emails, create marketing copy, elaborate and summarize content, write books, create music, or automate any repetitive tasks so that you focus on strategic tasks that require extreme human creativity and personal touch. It has amazing artistic skills that can be seen through its output stories, graphics, audio tracks, and videos. Although still in its early stages of development, generative AI has a huge potential to become a major force as this technology evolves. Some of the top trending AI tools in this category include ChatGPT, Midjourney, and DALL-E, which can be used to generate quality text, images, audio, and videos.

Explainable Artificial Intelligence

As the name suggests, explainable AI means the technology that can show evidence of its processes. There has been controversies with AI models that come up with ambiguous data without fully revealing how they generated and analyzed that data to produce their output. This led to questions of data reliability if humans cannot verify the processes of the machines in generating or processing data. Explainable AI is the technology gravitating more toward being responsible AI used for good intentions to eliminate any biases and hidden agendas by showing how it makes decisions. This can bring trust in areas where it is applied. For instance, healthcare and finance industries are sensitive sectors where reliability of data is a must. We can use data generated by explainable AI with confidence that it is not harmful nor misleading as other types with questionable output are.

Embedded Artificial Intelligence

Technically, embedded artificial intelligence (EAI), is a non-specific system for AI functions integrated in network devices with the aim to provide model management and obtain and preprocess data for AI-algorithm based functions in the same devices while offering low data transmission costs and ensuring data security. It is like having a smart helper built into your devices, like your phone or network equipment so that your devices can do clever things using AI without slowing them down. The main reason we need embedded AI is that this technology is changing the world really fast, and it needs three important things to work: Algorithms (smart rules), computing power (fast thinking), and data (information). But if each device had to have all of these, they would slow down a lot. EAI is like a superhero system that provides these three things to devices without making them sluggish. It achieves this by breaking down its tasks in the following manner:

Model Module
Model module is like the brainy part that has many different AI algorithms, like different ways of solving problems. You can choose which one to use for a particular task.

Data Module
Data module collects and gets the data ready for AI. It handles lots of data needed for AI tasks without overwhelming your device.

Computing Power Module

Think of computing power module as the muscle. It uses the algorithms from the model module and the data module to make quick decisions. Then, it sends these decisions to your device, which can use them to do smart things without needing to figure it all out on its own.

Currently, organizations are not only experimenting or integrating embedded in their systems but fully infusing the technology into their core solutions to make them more efficient, intelligent, and automatic. A great advantage of using embeddable AI is that it can be integrated into any device or system without the user having any extensive experience in coding. Additionally, its flexibility and ease of use allows it to be run on any device at the lowest cost, reducing development costs and enabling organizations to go to market faster.

Democratized Artificial Intelligence

Comparing the current AI systems with earlier developments, there is a trend of this technology being more democratized and accessible to everyone. With more open-source tools and platforms dominating the internet, anyone in the world can make use of AI, even if it is to a basic extent in underprivileged areas. Giant tech companies are also ensuring that we have democratized AI by providing a plethora of training, tutorials, and online courses so that these free tools reach the masses. Even though to get the best of the AI functionality often requires premium subscriptions, most developers aim to accommodate and reach as many people as possible by making the standard versions free to use. It is worth noting, however, that free AI tools are not all made accessible entirely for the benefit of the public. Data is the new currency, so essentially, we are willingly feeding these AI systems with our data, which the developers further use to train their models.

Multi-Modal Learning

Multi-modal AI indicates that the technology has really evolved into an interoperable scope rather than when it used to be operated in one dimension. As we saw with different types of weak AI, the previous technological era was mostly populated with task-based AI that specialized in certain functions only. However, the current trending tools are also able to multi-task and generate various modes of output from text, images, audio, and video. For instance, multi-modal AI can extract information from an image by understanding visual data as well as using the accompanying description text or caption. In healthcare application, multi-modal AI can combine data from medical images (MRI and CT scans) with data from patient records and help in the disease diagnosis or treatment planning. It can also be used in autonomous vehicles where it combines visual data from maps, camera sensors, image detection, and mobility to make wise driving decisions. Moreover, multi-modal AI can also be used to enhance customer service, as it can understand the customer's frustration by recognizing facial expressions and speech patterns to detect human emotions.

Computer Vision

Computer vision is the type of AI that enables the computer systems to see and understand the world around it. It is widely used in various visual-related applications such as facial recognition, self-driving cars, and medical image analysis. This technology excels more in increasing the speed of task-automation in any industry where major tasks are image-based. It can be further used to handle manufacturing tasks like 3D modeling, product counting, defect detection, inventory management.

Deepfakes

Having emerged in 2017, deepfakes are synthetic media generated through AI to portray false information. The term "deepfakes" is a portmanteau derived from deep learning technology used to create fake or non-existent content. Deepfakes are a common AI trend where images, audio, or videos are manipulated to depict events that never occurred or things that do not even exist in reality. It involves doctoring multimedia to look like someone said or did something they did not using tools such as face swapping. As you can imagine, deepfakes have the potential of being misused for malicious purposes such as framing innocent people or manipulating people's opinions based on fake news, scams, and propaganda. Indeed, deepfakes became a concern when the face of a famous actress, Emma Watson, was embedded into a pornographic video that went viral (Payne, 2023).

In 2019, the technology was also misused in a $243,000 worth scam where the voice of a UK-based energy firm's leader ordered the transaction to be made to a Hungarian supplier (Somers, 2020). It can further be used in this manner if scammers get hold of your communication channels and start contacting your family for emergency funds or fake your kidnapping and demand ransom.

When used for good, deepfakes would benefit the educational sector. Tests have been made where the technology was used to portray scientists explaining complex theories in simpler ways to understand and historical icons speaking their native language. Deepfakes can also be positively used in training AI models with more data of synthetic media of rare health conditions with less available material. Additionally, they can be used in language translation where a video is translated in multiple languages to expand a message for a good cause, as was seen with David Beckham's malaria awareness campaign (Payne, 2023).

Artificial Intelligence for Good

AI for good is the responsible and beneficial application of AI to solve social and environmental problems. It is a growing area of AI research, and it is being used in a variety of applications, such as fighting climate change and reducing poverty. In the words of most AI agents, including Sophia, they are designed for good and not for harm; therefore, AI for good can help us to create a more sustainable and equitable world.

Choosing the Right AI Tools

The internet is buzzing with a plethora of trending AI tools, so much so that it is hard to narrow them down and pick the ones you can use. In this section, I have categorized some of the top useful AI tools based on what they specialize in so that you can focus on those relevant to you. Whether you want to maximize your productivity, shorten your study workload, or create engaging content without spending chunks of time designing.

Productivity

Tools that optimize productivity generally function to help you get a lot done in half the time. Regardless of the complexity of the tasks at hand, having great AI tools ensures that your time is optimally used for tasks that would usually take long. For instance, manually creating a presentation used to be a time-consuming task of putting slides together, aligning your content, and still having to make it visually pleasing. However, with tools like Beautiful.ai, where you give the instruction to create a study or work presentation and the tool does it in less than a couple of minutes. You can drag and drop images and text to Beautiful.ai tool, tell it what you want it to do, and it generates the first draft that you can edit as you wish. You can also create your dream website in a matter of minutes with Hostinger AI website builder that requires no prior skills other than dragging and dropping your desired content.

Furthermore, you can use Descript to transcribe any video content to text in minutes. Instead of playing the video back and forth to extract its text, Descript transcribes any videos that come with subtitles. Powered by GPT-4, Taskade is another powerful tool that integrates notetaking, scheduling, and task management, helping users streamline their workflow and organize tasks effectively. As a Notion alternative, Taskade can also be used as a project management tool like the ones described below.

Chatbots

Chatbots bring efficiency to every industry and platform they are integrated into, as they ensure that customers are not left hanging while others are being serviced by the available agents. The ability to automate responses while still giving personalized interaction makes it easy for people to get help by describing their situations. This allows machines to handle numerous basic customer requests while leaving out only serious cases being escalated to be handled by humans.

Content Creation

Depending on the type of content you are creating, there are several tools you can use to design and create unique content. Make your imaginative art come alive by simply giving a text or speech prompt to an AI tool and watching it create for you. You can refine your images and videos by instructing the bot on different background colors, silhouettes, shades, and objects. Midjourney is one of the most incredible free-to-use AI tools that can generate graphics, book covers, banners, and profile pictures. While you pay nothing to use this tool, you must join Midjourney discord channel to access free editing tools. However, if you want privacy and exclusive ownership of your content, you can upgrade to a paid version. If your main focus is written content, check out the specific section below for incredible AI-powered tools that will take your short or long-form writing to the next level. Bing image creator is also another freely accessible AI tool to create or modify your art.

Project Management

Project management requires effort and cost efficiency, work and resource optimization, as well as error elimination. These are areas that can easily be amplified by integrating AI to automate some tasks in order to save time and reduce costs. Using AI can also minimize human error on tasks, enable easy assignment creation for teams, and track progress. Some of the best AI tools for project management include Notion, which is a free-to-use, versatile tool that manages a range of tasks from content creation, writing, editing, brainstorming, and note taking without switching across multiple platforms. It is a user-friendly platform that seamlessly fits into any workflow to streamline your tasks, update your teams, and collaborate on projects. It also comes at a paid premium price that unlocks extra features to optimize your work and enhance your productivity and creativity. Notion also edits and summarizes written content to improve quality.

Furthermore, Trello is another flexible and useful AI tool that takes project management to the next level. While this was just a project management software, its beginner-friendliness AI integration is a great additional tool that enhances work performance for beginners and expert project managers alike. With the ability to create boards, assign tasks, monitor progress, and optimize workflow, teams can stay on track with their project goals. It also has a great feature that enables you to integrate the calendar to ensure that everyone is aware of the deadline and is able to manage tasks in small chunks until the target is reached. Trello also offers a wide range of pricing options to fit into your budget. Other great AI-powered project management tools include Monday.com, Asana, Smartsheet, and Basecamp. These tools also offer varying price plans to accommodate small and large businesses that require streamlined workflow, focused teamwork, productivity, and efficiency.

Writing

If your focus is writing engaging content that touches hearts, it is a given that your readers prefer a more personal approach in your writing. The same is true if you write marketing pieces, SEO articles, landing pages, or copy. You want high-quality content that converts into sales. The application of AI in this niche had tongues wagging, as people thought AI-written content would be obvious and non-sentimental. However, that was before people knew the best ways to leverage AI tools like ChatGPT, WriteSonic, and Grammarly. From writing for you based on your prompts to helping you edit your written content, these tools save you a lot of trouble researching and organizing your content. As natural language processors, these AI tools understand prompts and are also able to deliver output in an engaging manner. You can even "teach" these tools to mimic your previous content so that they emulate the tone and deliver your desired message. You can ask these AI models to help you paraphrase, rewrite, summarize, or elaborate any content with any specific instruction (tone or audience-specific), and they can do it for you. However, you must be a responsible user and ensure that the content is factual and not plagiarized. You must also ensure that your content sounds human-written by using AI detection tools like Originality.ai, Duplichecker AI detector, and ZeroGPT.

Studying

Studying can be challenging for most people with learning limitations such as dyslexia. This is the area where AI is getting affirmative thumbs up from both learners and educators. Some of the best tools you can use include Speechify, a tool that converts any written text to spoken word so that learners can listen to the material instead of reading it. This improves comprehension and memory retention. Speechify has a great feature where users can regulate the playback speed as they desire; you can increase the speed to improve your reading speed or reduce it to pay attention to detail. You can change accents, languages, and even customize celebrity voices for fun learning.

Furthermore, Humanata AI is one of the great AI tools where you can upload your school notes and textbooks and then ask the bot any question related to the uploaded material. Not only will it give you the correct answer, but it will include the page number for reference and a summarized or lengthened version of the text.

For research and academic studies, Quillbot is an awesome AI tool with built-in paraphrasing, translating, grammar and plagiarism checking, citation generating, and summarizing features. It allows you to modify the tone of the content without losing the core meaning, ensuring that it remains authentic. Adding to this research-focused studies, Genei is an incredible tool that extracts and summarizes the key notes from any research material, enables you to build your own research library, and organizes everything into easy to navigate notes. It also helps expand or shorten your content as you wish to give you enough information without overwhelming you with an entire article. Like Quillbot,

Genei is also able to generate citations for your content. It comes at $9.99 per month with 40% discount for academic students (Dr. Alex Young, 2023).

Chapter 4: Ethical Challenges of AI

Along with the rise of the technological advancement are increasing concerns regarding AI ethics. AI ethics essentially mean the responsible and moral application of the technology by adhering to the guidelines, principles, and standards governing its development and deployment. The responsible and moral use of AI means considering the ethical implications and societal impacts of AI technologies and ensuring that they are developed and applied in ways that align with human values and respect fundamental human rights. It is paramount that AI systems must be fair and not discriminate against individuals or groups based on race, gender, ethnicity, or other factors. If there are any biases in data and algorithms, AI ethics would require them to be addressed. Rather than being super cryptic by hiding methods and strategies of data processing, there must be transparency, making AI systems more understandable and explainable so that users can have insights into how AI decisions are made. AI ethics should be everyone's concern because even those who are not interested in the technology may be victims of AI-driven decisions directly affecting their lives. From what determines interest rates on loans or loan approvals to college admissions or employment outcomes, if AI algorithms are used in these decisions, there are likely to be disparities that most people are not even aware of. Therefore, this chapter succinctly addresses AI ethics in detail and points in the direction society must go to ensure that ethical challenges are dealt with.

Ethical Concerns Voiced by Tech Industry Leaders

As I mentioned, most AI users and non-users are oblivious to the AI ethical conundrums; therefore, it is usually those who have been in this space for longer who are trying to educate the masses about AI ethics and their code of conduct. Having used or witnessed AI applications firsthand, tech industry leaders have the experience of the limitations of AI and are on a mission to make this space safe for everyone. This section is a summary of some of the major ethical concerns voiced by those who understand the technology better and want to improve its application in society.

Privacy Violation

AI systems are trained on massive datasets of personal data, which raises concerns about privacy and security. If this data is not properly protected, it could be hacked or stolen and used to harm individuals or society as a whole. For example, AI could be used to create deepfakes that could be used to blackmail or extort people.

Bias

AI systems learn from the data they are trained on, which means that they can reflect the biases that exist in that data. For example, if an AI system is trained on a dataset of images that are mostly white and male, it may be more likely to identify white males as faces. This can lead to unfair outcomes, such as people of color being less likely to be hired for a job or being stopped by the police.

Lack of Transparency

AI systems can be complex and opaque, making it difficult for people to understand how they work and why they make the decisions that they do.

Lack of Emotional Intelligence

As I mentioned earlier, AI systems do not have emotional intelligence, which means they cannot understand or empathize with human emotions. This can make it difficult for AI systems to make decisions that are fair and just. For example, an AI system that is used to decide who gets a loan may not be able to take into account the borrower's personal circumstances, such as whether they have recently lost a job or experienced a medical emergency. They are often rational in their analysis, leaving no room for humanity or compassion.

Hindrance of Societal Advancement

Overreliance on AI to make groundbreaking discoveries and research prohibits people from influencing change or challenging the status quo. The thought of being outsmarted by a robot can make people hesitant to bring new ideas to the table in case they are not regarded as smart. Even when someone sees a different angle that counteracts an idea brought up through an AI system, they might not be as willing to share it with others. This notion hinders societal advancement, as people would rather be quiet than voice their opinions.

Limit Human Willingness to Make Difficult Choices

The ability for AI to make fast and informed decisions where humans would naturally be hesitant may be good for company efficiency. The question is, at what cost? The more people know that AI will make hard decisions for them, the more reluctant they will be to think outside of the box or even pioneer new theories. This kills boldness, self-confidence, and control over situations that require human leadership. If perpetuated, this will lead to timid people who do not trust their own capabilities to lead and influence others.

Unfair Outcomes

Since AI systems are trained on data that is collected from the real world, they can reflect the biases, prejudices, and inequalities that already exist in society. For example, an AI system that is used to predict who is likely to commit a crime may be more likely to predict that people of color are likely to commit a crime. This can lead to unfair outcomes, such as people of color being more likely to be incarcerated.

Harmful Events Perpetuated by AI's False Predictions

AI systems can make mistakes that can have harmful consequences. For example, an AI system that is used to diagnose diseases could misdiagnose a patient, leading to the patient receiving the wrong treatment. AI systems could also be used to create fake news or propaganda, which could mislead people, ruin reputations, and lead to harmful events.

Another instance can be if AI systems are used to measure the timing and severance of an upcoming natural disaster like a cyclone or an earthquake. Miscalculations can lead to residents not evacuating the targeted place on time or risking their lives and belongings that could have been spared if the prediction was accurate.

Threat to Job Security

As much as most people love having their workload lifted and the boring tasks handed over to AI, automation is a major threat to people who solely know and enjoy doing the mundane tasks. For instance, most companies are now entrusting the majority of their admin work and routine inquiries to chatbots. This has caused a significant drop in the need for human agents. While some companies are trying to avoid completely leaving other people out of jobs, creating other positions to oversight and monitor chats or completely diverting into new positions, it is also becoming expensive to train everyone for other roles. No company is able to bring everyone along for the joyride, as some people are older and less keen to learn new skills for their new roles. Now, replacing human resources with impersonal technology calls for ethical concerns, as hiring a machine is different from hiring a human being who has dependents and a social role to play.

A Thrill Sport for Criminals

With cybercriminals equally having free access to AI-powered tools, there is a rising danger to humankind and a greater threat to cybersecurity. One unavoidable thing about criminals is the rate at which they adopt and misuse technology for their own advantage. Remember how the early adopters and users of blockchain technology, particularly Bitcoin, were people operating in the dark web? The idea that a technology created to improve the monetary system by reducing currency inflation is used for illegal activities taints the good that it represents. Additionally, its anonymous nature and freedom to be created by anyone also enables bad actors to continue swindling ignorant people's funds. This leads to the slow adoption for greater purpose of cryptocurrencies and calls for strict regulatory measures taken against technological advancement. The application of AI in blockchain and other industries continues to leave a bitter taste in the mouths of crime victims as it makes it harder to trace and punish bad actors. For instance, people still use AI bots to infiltrate presales, front-run transactions, and manipulate prices, resulting in unfair project launches.

Moreover, I have already mentioned in the previous chapter that deepfakes are another AI application misused for harmful activities. People's reputations are at constant risk as scammers continue to manipulate data and the victim's loved ones become the target for further exploitation. All these negative sentiments and criminal activities surrounding the technology make it harder for people to trust that AI can be used for a greater purpose without benefiting criminal activities. It is no surprise that industry leaders continue to warn us that AI in the wrong hands is catastrophic.

Privacy Concerns and Data Ethics in AI

Privacy concerns and data ethics in AI are critical aspects of AI ethics that focus on protecting individuals' privacy rights and ensuring that data used in AI systems is handled responsibly and ethically. Addressing the following privacy concerns and adhering to data ethics principles involves clear policies, transparent practices, informed consent, robust security measures, and ongoing oversight to ensure that AI technologies respect individuals' privacy and data rights while delivering their intended benefits.

Data Collection and Consent

AI systems often require access to large amounts of data, which can include personal information. Collecting this data without informed consent can infringe on individuals' privacy. For example, a mobile app that collects location data without clearly informing users or obtaining their consent is infringing on their privacy.

Data Security

The concern here is that inadequate data security measures can lead to data breaches and unauthorized access to sensitive information, posing significant privacy risks. For instance, a healthcare AI system storing patient records without robust encryption, making them vulnerable to hacking, is a significant security risk. With AI systems, gone are the days of doctor-patient confidentiality, as data that was supposed to be confidential is now at risk of being accessed by bad actors and used without the owner's consent.

Data Anonymization

Re-identifying individuals from supposedly anonymized data can compromise privacy. AI techniques can, sometimes, de-anonymize data, revealing personal details. This is evidenced by researchers successfully re-identifying individuals from supposedly anonymized medical records.

Bias and Discrimination

As discussed earlier, biased data or algorithms can result in discriminatory AI outcomes, which can disproportionately affect certain groups, raising ethical and privacy concerns. For example, an AI-powered recruiting tool that discriminates against job applicants based on gender or race due to being trained on biased or discriminatory data will result in questionable outcomes.

Surveillance and Tracking

AI-powered surveillance systems can infringe on privacy rights by tracking individuals without their knowledge or consent. The example can be seen with the widespread use of facial recognition technology in public spaces for surveillance purposes.

Data Ownership and Control

Clarifying who owns and controls data used in AI systems is an ethical consideration, as individuals may want more control over their data. For instance, there may be questions about who owns the data generated by a smart home device; the user or the device manufacturer?

Ethical Data Sourcing

There are currently concerns about the ethical sourcing of data for training AI models, especially when data is collected from social media or online sources. Getting already-exposed information means that people have no control over what is made public and what remains private concerning their lives. Therefore, ethical sourcing of data includes ensuring that data used in AI is acquired legally and ethically, without infringing on privacy rights.

Addressing Bias in Artificial Intelligence Algorithms

Bias is one of the primary ethical concerns when dealing with AI-generated data. Because AI systems are trained with existing data, they often tend to use the same data with its limitations and preexisting bias. AI bias needs to be addressed to ensure fairness, inclusivity, and lack of prejudice. Addressing AI bias is a responsible thing to do to ensure that AI-generated data is transparent and that proper guidelines are adhered to during data collection, processing, and analyzing. This will lead to humans trusting more in the system and its outcomes without having doubts or questions of any hidden or skewed data.

In What Way Is AI Biased?

Common bias issues in society that AI algorithms produce include racial, sexism, and disability discrimination. If an AI is being used to select employment candidates, it might select more men over women for high paying jobs. If it is used by credit unions and banks, there is a likelihood of people of color being denied loans because they are deemed as high-risk or likely unable to repay their debt. Additionally, there have been several reports on the disparities of AI misidentifying people based on race in a harmful way. For instance, increasing the likelihood of people of color to be the prime suspects in criminal activities. As more young people quickly adopt the technology, the elderly are usually left on the sidelines and helpless, as it takes serious patience to convince them to fiddle with the technology until they are confident. To avoid being a burden to others, those who feel AI is too complex for them often socially isolate themselves. Rather than curbing the problem, by bringing more courses to those who are not tech-savvy, AI increases this social isolation as things continue to become more easier for the younger generation and more complicated for those who are slow to take any action.

How Can Bias Be Addressed?

Transparency and Explainability

There must be transparency on how AI algorithms collect and analyze data. AI systems must also reveal their methods and strategies so that they can be verified for quality assurance and fairness.

Fairness

People must be more vigilant on their systems to ensure that the outcomes are fair and free from preexisting bias issues. Organizations must also be more flexible to adjust their systems if there are any signs of unfairness.

Diversify the AI Field

Instead of focusing on narrow data collection and processing, there must be more investments toward a diversified AI field that uses a wide range of demographics and data collection procedures.

User Data Rights

Whenever personal data is accessed, people must be left with the choice on how their data is used or shared across different platforms. This means that data should not be recycled but collected fresh every time as needed. For instance, statistics for people of color out of jobs must not be used to determine another numeric-based research involving people of color. Those who own data must also have rights on how it is distributed and used to prevent re-using data that contain bias for a new search.

Proper Guidelines and Procedures

There must be proper ethical guidelines, auditing techniques, and tools must be deployed to minimize biases. The European Union (EU), Federal Trade Commission (FTC), Equal Employee Opportunity Commission (EEOC) are some of the regulatory bodies actively dealing with AI bias. These ethics-focused organizations are ensuring that AI systems, developers, and companies actively using AI are accountable for their practices and use of the technology for the good of society.

Chapter 5: The Role of Humans in the Age of AI

Most people who choose to ignore any technological advancements think by not participating, they are not supporting them. However, that equates to being a non-voter; just because people are not actively electing their leader does not mean they are not picking any sides. Instead, they are simply choosing to have others make decisions for them. If people who are anti-technology remain silent and passive, that will not stop its evolution. Rather than standing on the sidelines like humans who have no control over the evolution and advancement of these machines, it is time to thoroughly explore the human-machine partnership. I believe that knowing what role to play in this era will ensure that humans maintain the upper hand over artificial intelligence and its unpredictable future.

Developing and Designing AI Systems

As creators and users of AI machines, it is our responsibility to ensure that this technology serves, rather than threatens, humanity. This starts from regulating the intention of creating AI systems and ensuring that proper procedures are employed before approving development and introduction of various AI systems to society. The rapid growth of AI requires people who understand the impact of the technology advocating for its responsible use from the initial design. Innovators must be held accountable to educate society on their creations. From the skeletal framework of AI agents to their intended updates, there must be transparency of the design, function, and purpose of a created AI system. If innovators and policy makers can reach consensus and openly agree that every AI tool entering the market can be traced back to its blueprint, then we can rest assured that we live in the era of transparent AI intended for good.

Using the Technology to Improve Our Lives

In the age where AI is overhyped, it is possible for governments and organizations to want to hop-on the bandwagon due to the fear of missing out. However, it is crucial for those in power to regulate these processes and ensure that AI systems are designed to improve lives and are, thus, executed with that purpose. It is possible that some people develop these systems for personal gain and entertainment without worrying about their repercussions. To reiterate, AI in the wrong hands can be detrimental to society. Which is why there must be a clear roadmap and regular audits indicating that AI is being used to improve our lives. Companies in the forefront of technology must give detailed reports of their AI usage and showcase their progress in terms of the number of issues solved through AI. If an organization claims to assist society in any way, there must be accompanying reports to support their claims and regular audits to ensure that there is no foul play or data manipulation. There must also be strict regulations on people who develop AI models without solving real-world problems to avoid using the technology with ulterior motives.

Governing the Use of AI

There must be strict policies governing the use of AI. However, the reason behind the slow implementation of regulatory processes is that governments are often lagging behind innovators, giving them the first mover's advantage over how these policies are put in place. Governments and regulatory bodies often struggle to keep pace with the rapid development of AI technologies. Innovators and tech companies can move quickly, creating and deploying new AI systems before comprehensive regulations can be established. For example, autonomous vehicles have been on the roads for years before some governments have developed clear regulations for their safe use.

Organizations that are early adopters and developers of AI technologies may gain a competitive advantage before regulations catch up. They might influence policy development to align with their interests. Tech companies that develop AI-powered data analytics tools might use customer data for profit before data privacy regulations are enacted. Moreover, the main challenges over regulations include the rate at which AI systems are developed versus the time these issues are discussed in relevant regulatory bodies. The pace at which AI systems are developed can significantly outstrip the time it takes for regulatory discussions to address emerging issues. This time gap can lead to unaddressed ethical and safety concerns.

Another challenge is that AI is multifaceted, making it difficult to govern what needs to be regulated without generalization. It is impossible to overrule AI applications as they benefit different industries at different levels. As previously discussed, AI applications range from autonomous vehicles and healthcare diagnostics to financial algorithms and content recommendation systems, each requiring unique regulatory considerations. Even if policy makers wanted to implement a ban on AI on industrial level, what about in areas where AI application is the best solution? Can we afford to overlook all the good when we ban the technology? The question that always remains is how to strike a balance between adopting technological innovations and protecting society. Striking the right balance between fostering innovation and ensuring responsible use of AI is challenging. Overly strict regulations can stifle technological advancements, while lax regulations can lead to ethical and safety concerns. This requires neutral advocates to represent both the innovators and the public to ensure that there is a balance of common interests.

Furthermore, AI operates on a global scale, and yet regulatory standards and approaches can vary widely between countries and regions. The absence of a global consensus can create challenges for international cooperation and harmonization of regulations. For example, privacy regulations in the European Union, like the general data protection regulation (EU GDPR), differ from those in the United States, creating compliance challenges for multinational tech companies. These challenges highlight the need for proactive and adaptive regulatory frameworks that can keep pace with AI advancements while considering the multifaceted nature of AI applications. Policymakers, industry experts, and stakeholders must work collaboratively to develop policies that promote innovation while addressing ethical, safety, and societal concerns associated with AI.

Protecting Ourselves and Others From the Risks of Artificial Intelligence

As you are reading this book, you are already getting informed of the impact of AI on society. Your role does not end with you benefiting from this content in your personal capacity only. What about those around you who are oblivious to this technology? It is your responsibility to educate them so that they also know the benefits and risks of AI. They may be victims of data privacy violations or other AI biases without even knowing. Opening their eyes helps them to make informed decisions.

Explore Open-Source Artificial Intelligence Platforms With Caution

It can be exciting to sign up and freely access AI tools on trending platforms. However, be cautious of your information because these platforms retain your data and use it to further train their models. Samsung employees accidentally leaked critical information while using ChatGPT at work. One employee revealed trade secrets when checking the source code for errors while another shared a recording of a meeting to transcribe it into notes for a presentation. This type of information in the wrong hands is detrimental to the company (Mauran, 2023). You can even consider creating accounts with a fake alias to protect your information on these platforms.

Be Selective About What You Share Online

Most people display every detail about their lives online, including sensitive information about their family, location, and what they plan to do. AI can be used to extract your habits, beliefs, and preferences from your shared information. Even if people do not interact with your shared data, you cannot be sure how many people are monitoring you or what they are planning to do with your information.

Do Not Save Your Login Details on Online Platforms

I know the simplicity of one-click login on any frequently visited online platforms; however, you must avoid saving your login details and always input them afresh. Scammers are always phishing for easily accessible accounts, so avoid this vulnerability.

Use Strong Passwords

Avoid using easily predictable passwords like birthdays or names of loved ones on any online profile. You must use long passwords that have a combination of different characters and symbols for a strong sequence. You must also enable two-factor authentication or one-time pin codes to double your online security wherever applicable.

Do Not Ignore Potential Bias

We have already established that bias exists in AI algorithms; therefore, you must always be vigilant to ensure that proper standards are adhered to wherever your data is used. It is your responsibility to ensure that there is fairness concerning your data, or that of your loved ones, wherever applicable. If you applied for a job or a loan, you must seek to know how the application process goes and if the organization explored enough data before giving an outcome.

Ensuring Responsible Use of Artificial Intelligence

Most industry experts monitoring and exploring the evolution of AI often mention that the technology itself is not that scary or harmful to mankind. Their biggest fear is those working behind the scenes to bring these machines to life and do amazing things. It is paramount that you stick to the responsible use of AI and use the time you saved by delegating boring tasks to enhance your quality of life.

Obtain Informed Consent From Users

No one should ever find themselves using AI without understanding what they got themselves into. Wherever AI is applied, people must be informed and be allowed to give consent instead of being rushed to make a decision without a basic understanding. This means AI systems must be applied fairly to a community that truly understand the purpose of AI use and to what procedures are emplaced. If people's data will be shared, they must be informed and in agreement with this. Anyone not giving consent must also be respected instead of being maltreated or blamed for dragging down the progress of an organization.

Freedom to Opt In or Out

No matter the size of an organization, employees and customers must never be coerced into participating in AI-powered programs. People must be given a choice to get involved, even if on a trial basis only, and to exit if they feel like the program is not in alignment with their values. They should be allowed to freely and easily withdraw without having to explain themselves. This means that those who implement these programs are not forcing anyone into doing anything they are uncomfortable to.

Debiased and Explainable AI

As briefly highlighted in the previous chapter, all applicable AI usage must be explicitly and transparently explained and articulated on how biases are mitigated in decision-making processes. For instance, if a company used AI-algorithms to make any hiring decisions, they must prove that the data used contained no biases and that there was inclusivity and fairness.

Transparency on Data Privacy and Involvement of Third-Parties

Organizations must be transparent with how they collect and use people's personal information. They must provide detailed measures taken to protect privately stored data. If they share data with third-parties, they must also be transparent and give people a choice to agree or decline these terms.

Continuous Learning and Training

People must be exposed to continuous learning and training in the technology, as it is rapidly changing. It is unwise to stagnate with basic understanding of how AI works because what we know today is different from what was known in the past. Similarly, the information we hold dear today may be irrelevant in the future. To keep up with the lightning speed of AI evolution, there must be constant classes and programs that keep society informed.

Health and Wellbeing Must Be Prioritized

There must be regular checkups and monitoring to ensure that the continued use of AI tools and programs at workplaces and other areas where people are exposed do not negatively impact people's health and wellbeing. It is possible some workers may feel overwhelmed or stressed as they familiarize themselves with the new system. Therefore, health and wellbeing must be prioritized at all times.

Chapter 6: The Future of AI

Over 133 million new jobs are expected to be created in relation to AI by 2030 (Matleena, 2023). One of the major fears has been that AI will replace jobs and create a huge unemployment gap. However, it is also expected that there will be the creation of more fascinating roles for humans willing to stay in the game. As conversations on the burgeoning adaptation AI technology continue to dominate major platforms, the million-dollar question remains: What does the future of AI look like? Privacy issues, strict regulations, speed of change, and AI-human partnerships are some of the current issues that are also futuristic. Some experts are calling for a pause in the innovation of AI and lab experiments to ensure their safety and positivity toward humanity. The exponential rate of AI revolution and adoption calls for attention and vigilance, as the future looks nearer than we anticipate.

What Does the Future Hold for Artificial Intelligence?

Artificial intelligence is no longer confined to the realm of science fiction; it is very much a part of our present and an integral part of our future. This dynamic technology is not just going mainstream; it is reshaping the way we live, work, and interact with the world around us. And as AI continues to advance at breakneck speed, it is keeping everyone on their toes, from tech enthusiasts to policymakers and everyone in between. One of the most striking aspects of AI's future is its relentless expansion and diversification. As AI becomes increasingly accessible and its benefits become more apparent, we can expect to see it branching out into specialized platforms and domains. From healthcare to finance, education to entertainment, AI is poised to revolutionize every facet of our lives. The possibilities are limited only by our imagination and our ability to harness this technology for the greater good.

However, amid the excitement and optimism, there are also apprehensions. The notion of self-aware AI and conscious machines looms as a thought-provoking, but somewhat unsettling, possibility. While it is a topic of debate, it is important to recognize that we are still far from realizing this level of AI sophistication. Some scientists and influencers may argue that this is fast approaching; however, no one really knows the timeline or even the possibility. So, at this point, this is currently mere speculation and fear of uncertainty being instilled in people. AI today is more about narrow or specialized intelligence, excelling at specific tasks but lacking the general intelligence and self-awareness associated with human consciousness. We can merely keep watch of this space to see this futuristic advancement occur or stay engaged in developments toward it so that we can take action based on verifiable information.

Moreover, the future of AI is also closely tied to societal choices. As AI becomes more ingrained in our lives, we are likely going to witness a continuous wave of mass adoption. Those who embrace AI will find innovative solutions, enhanced efficiencies, and new opportunities. For them, AI will be a transformative force that empowers them to achieve more. Contrarily, the rise of AI also brings about concerns, particularly regarding ethics and responsibility. We can anticipate more stringent policies and regulations aimed at ensuring that AI is developed and used responsibly. Privacy, fairness, transparency, and accountability will be at the forefront of these efforts. Some may even resist AI's pervasive influence, advocating for strict regulations or resisting its integration into their lives.

In this ever-evolving landscape, staying engaged in the future development of AI is not just advisable; it is essential. It is a call to action for individuals, organizations, and societies to actively participate in shaping the AI-driven future. By remaining informed, we can refine our strategies, make informed choices, and ensure that AI aligns with our values and serves the best interests of humanity. In this way, we can navigate the exciting, yet complex, journey into the age of artificial intelligence.

The Future of Jobs in the Age of Artificial Intelligence

Imagine a world where your colleague is not a person but a bot, tirelessly crunching numbers and analyzing data without a hint of fatigue. This is the future of jobs in the age of AI. While AI is eliminating some traditional roles, it is also the architect of a job revolution. Arguably, the future of jobs seems to be more fascinating than scary for optimistic people who are up for a challenge.

Eliminating Some Jobs

AI is not just about automating tasks; it is also about transforming industries. Jobs like data entry, routine data analysis, separating things, picking fruits, answering customer service calls, and even some customer service admin roles will become obsolete for humans and entrusted to AI. AI can handle more than double the capacity, so eventually, organizations will realize that replacing humans with AI for these positions saves them time and money. As previously mentioned, unless people upskill themselves, employees specializing in these automatable jobs will face retrenchment.

Creating New Jobs in the Process

As AI takes over routine tasks, it frees up humans to take on new roles. Think AI trainers, prompt engineers, data scientists, and AI ethicists. These are the jobs of today. As technology advances, tomorrow we might be exploring new skills to keep up and stay employable.

Changing the Nature of Work

As AI changes the nature of our work, it is also increasing the demand for skills in science, technology, engineering, and mathematics (STEM) fields. Suddenly, coding is the new cool, and data literacy is an intriguing subject. Even outside of STEM fields, almost everyone is now interested to know how AI can be applied in their line of work. Gone are the days when it was considered as a field of study for STEM nerds; anyone, regardless of their field of study, is now able to interact and integrate AI in their workplace.

The Impact of Artificial Intelligence on Society

Now, let us talk about how AI is shaking the very foundations of our society, challenging norms, and raising questions we have never faced before. We have already seen the positive impact of the technology in the industries where it is applied. While improving efficiency is one of the core upsides for AI, there are other areas of our lives where we should be concerned about how AI affects them.

Changing the Way We Interact With Each Other

Thanks to AI, we chat with bots like they are old pals. We no longer have to stand in long lines to be assisted with administration, as most organizations integrate AI in their systems and online platform. However, this is not just about chatbots, but also about redefining how we communicate with both humans and machines. We now know that we have to input a clear prompt for the best output. If we can do this in our communication with machines, how much of an improvement will this have on human relations? AI is changing the way we communicate, we must be careful to ensure that while we improve human-machine relationships, our human-human relationships are not neglected but get more attention as AI frees up some of our time.

Challenging Our Beliefs

AI is throwing curveballs at our beliefs. It can predict our preferences; sometimes, even better than we know ourselves. With synthetic media dominating our sources of information, making us question what we think we know, soon the truth will be blurred. Even though we are more curious to know and explore the things we previously thought were humanly impossible, deepfakes will challenge our beliefs as we encounter distorted truth or fake news about beliefs and values we hold with high esteem.

Redefining Our Values

AI can discern patterns in data that humans cannot and spend less time making decisions while humans would usually ponder on information and view it from different angles before making any rushed decisions. This means AI is challenging our values and ethics, pushing us to reconsider how we make decisions. The rate at which decisions will be made will be fast, affecting the current values of verifying data and ensuring that it is not harmful to society. Some people might even value quantity of work done over quality in the name of being overly efficient.

Creating New Social Norms

Remember when talking to your phone hands-free in public seemed weird? Now, it is normal. AI is shaping new social norms, and it is only the beginning. Some gatherings for information briefings might not be necessary in the future as communication becomes more automated. This will reduce human interaction and increase the amount of time people spend connecting with machines rather than with each other. The rise in social robots is an evident example that people will find companions in machines, diminishing the quality of friendships and human relationships.

Where Does That Leave You?

In the realms of an unpredictable AI future, what can you do to stay ahead of the game? Well, maybe you are not interested in taking a front row place as the technology advances. However, you may want to remain engaged because at the rate at which AI is evolving, you might wake up and find yourself left far behind, and it becomes a struggle to find a relevant role to play in society. This is why you must at least know a bit about how the developments are going as well as the regulatory progress.

Integrating Artificial Intelligence in Your Everyday Life

From already using your digital personal assistant to controlling your smart home, you are at the helm of AI's integration into daily life. Start small, get comfortable, and let AI be your assistant as you explore its depths. The more you have fun with it exploring new features, the easier it will be to integrate more complex AI applications when necessary.

Deciding How Much Personal Information to Share

You have already learned that AI loves and thrives on data. Therefore, you must decide how much you want to share going forward. Be cautious, protect your privacy and understand the trade-offs. There is always something you are giving away the more you share your data on AI platforms. Be vigilant and restrictive to know when you have shared enough to get your desired tasks done while still protecting yourself.

Being the Responsible User of Artificial Intelligence

In this age of AI, being informed and engaged is your superpower. Do not stay on the sidelines and let other people decide for you. Keep up with AI developments and upskill when needed. Stay curious and adapt to the ever-changing landscape. Protect your loved ones by ensuring that they also know the responsible ways to use AI while protecting their privacy.

Advocating for Ethical Artificial Intelligence Practices

As we discussed earlier, you have a role to play in ensuring that proper practices are employed as we embrace the technology. Therefore, you must consider joining the movement for responsible and ethical AI. Be a part of organizations that promote AI development with human values at its core. Your voice matters in shaping AI's future. You can also find legitimate organizations to support if you are in a position to financially support ethical missions to keep this space safe.

Monitor Your Family's Involvement With Artificial Intelligence

Just as it is important to limit your children's time spent on digital gadgets, it is your responsibility to ensure that you maintain a family unit that connects with each other. Even if you use AI to automate some of the tasks within your household, you must ensure that your family still makes time to actively engage in manual activities and physical connections. This will also contribute to great health practices instead of creating lazy family members who press a button every time a chore awaits. Additionally, you must also monitor that your children combine natural learning practices that improve their thinking capacity rather than solely relying on automated and enhanced study materials.

Conclusion

That was a quick ride into technological space, wasn't it? From exploring the meaning of artificial intelligence to demystifying all the technical jargon to make this an interesting and engaging read, as well as looking into the applications of AI and your role in this evolutionary topic, I absolutely loved embarking on this journey alongside you. Thank you for dedicating your time to navigate the transformational world of AI instead of watching as our world becomes reeled into unfamiliar landscapes that will only feel more foreign the more it changes right in front of our eyes. You have already realized that there is no stopping this technological advancement, as more people are opening up to the idea, adopting the applications, and investing in innovations to expand the AI landscape. You being here, imbibing this information, is the best thing you can do to remain engaged and enlightened about our current world and what it can potentially be.

To recap, I started by explaining the concept of AI in simple terms as the training of computers and machines to be able to think, learn, understand, and explore data in a manner similar to how humans learn, interpret, and interact with information. From their inception, machines were designed to resemble our cognitive abilities of analyzing data to find meaning and understanding as well as be able to use that information in the future. I spoke about the structure of neural networks being inspired by the layering of the human brain to transmit information. As the components of deep learning, which is a subset of machine learning and AI, neural networks are designed to learn and make predictions based on the data they have been trained on to solve complex problems. All these subfields of AI display the ability to read and recognize data the same way humans identify images, hear sounds, and recognize patterns to use that information the next time we interact with that data or find new information. These advancements preceded what we later saw under the incredible applications of AI in various sectors like healthcare, the financial industry, enhanced education, arts and entertainment, mobility and transport, as well as cybersecurity.

Furthermore, we time traveled into the past to witness the evolutionary journey of how AI was conceived. From programming machines to solve complex math problems, play chess, and translate languages, the AI evolution has been a rollercoaster ride filled with ebbs and flows. After the coining of the term "artificial intelligence" in 1950s at the Dartmouth College conference, more research and funding went into this field of study, with pioneers discovering some interesting concepts in the process. There has been phenomenal growth in expert systems, natural language processing, machine learning, neural networks, and the machines' ability to recognize speech and advance visually. However, the major development in AI was witnessed in robotics, as artificial bodies evolved from having basic mobility to being able to carry out tasks like moving items from one place to another and assembling things. Robotics also made headline news when humanoid robots not only displayed mobility functions but human-like intelligence in processing information, reasoning, and transmitting data. Although there was an AI winter—a period when people called doom on the concept and innovations halted—the successive surge of the technology grew exponentially until the current day, where we do not even seem to be in full control of where it is headed.

Arguably, the advancement of AI has been interesting to watch, even though it seemed pretty complex from outside. However, since the current frenzy that can be attributed to generative AI and the free access to OpenAI's brainchildren, GPT and DALL-E, most people saw the need to take a more hands-on approach and learn by doing. While some tech industry leaders tried to warn us about the negative implications of AI, if not handled well, not many people paid attention then because the topic was beyond their comprehension. Additionally, some people also thought that by doing nothing, they were exempt from the effects of AI, oblivious that they may be directly implicated even if they were not aware. The education around AI ethics is one of the most groundbreaking components of making this technological space safe for all of us. Addressing the AI algorithm bias, unfairness, data privacy violation, lack of transparency, the impending high unemployment rate resulting from AI replacing human jobs should be everyone's concern. There needs to be an active involvement by policymakers and the public to ensure that AI applications are strictly aimed at improving humans' quality of life instead of threatening it.

My main purpose with *Artificial Intelligence Pushing Boundaries, Is It Worth It?* was not to give you an arsenal of tools to understand this technological space, and then you fold your arms and do nothing. It is an invitation to be a fellow responsible partaker of AI who has a basic understanding of what it entails. You have seen that, from the experts' point of view, the future looks scary not because machines seem dangerous, but because AI in the wrong hands is a ticking bomb that might explode before us. My invitation is for you to stay informed and active as this technology rapidly advances so that you can be in a position to advocate for ethical and responsible applications of AI for the good of humanity. Together, we can create a firm wall of protection even if we see some of the developers using the technology for harm. Let us ensure that our loved ones are also aware of this behemoth concept and hope they can support the advocacy for its responsible societal improvement.

In a nutshell, after going through everything contained in this book the responsibility is left with you, dear reader, about what you will do with this information. My job was to bring the subject closer in a natural language we both understand instead of trying to influence your decision with one-sided information. I believe that this book delved into both the exciting, great benefits and bone-chilling risks and downsides of AI so that we weigh what we can improve on and what to subtract going forward. If you have found this book as helpful as I intended, I will greatly appreciate you leaving a review on Amazon so that more people are exposed to this information. By so doing, you will have started on the joint positive work to make this technological space safe for the future of humanity.

Glossary

Algorithm: A step-by-step procedure for solving a problem.

Artificial Narrow Intelligence (ANI): A type of AI that is focused on a single task or set of tasks like playing chess or identifying objects in images.

Artificial General Intelligence (AGI): A hypothetical type of AI that would be as intelligent as a human being and be able to understand and perform any task that a human can.

Artificial Superintelligence (ASI): A hypothetical type of AI that would have a level of intelligence far greater than that of a human. ASI would be able to solve problems that are currently intractable for humans.

Augmented Reality (AR): A technology that overlays digital information onto the real world. AR can be used for gaming, navigation, and education.

Bias: A prejudice that can skew the results of an AI system. Bias can be introduced into AI systems during the training process, when the data that the system is trained on is biased.

Big Data: Large amounts of data that are difficult to process using traditional methods. Big data can be used to train AI systems and make predictions.

Blockchain: A distributed ledger technology that can be used to record transactions in a secure and transparent way.

Chatbots: Computer programs that can simulate conversation with humans. Chatbots can be used for customer service, education, and entertainment.

Cloud Computing: The delivery of computing services over the internet. Cloud computing can be used to train and deploy AI systems.

Computational Power: The ability of a computer to perform calculations. Computational power is important for training and deploying AI systems.

Computer Vision: A field of AI that deals with the recognition and interpretation of images and videos.

Convolutional Neural Networks (CNNs): A type of neural network that is well-suited for image recognition tasks. CNNs are used in many AI systems, such as facial recognition and self-driving cars.

Cryptocurrencies: Digital or virtual tokens that use cryptography to secure their transactions and control the creation of new units. Cryptocurrencies are decentralized, meaning they are not subject to government or financial institutional control.

Cybersecurity: The practice of protecting computer systems and networks from attack. Cybersecurity is important for protecting AI systems from cyberattacks.

Deepfakes: Synthetic media in which a person in an existing image or video is replaced with someone else's likeness. Deepfakes can be used to create fake news videos or impersonate people.

Deep Learning: A type of machine learning that uses artificial neural networks to learn from data. Deep learning is used in many AI systems, such as image recognition and natural language processing.

Embedded Artificial Intelligence (Embedded AI): AI that is integrated into devices, such as smartphones and cars. Embedded AI can be used to improve the performance and functionality of devices.

Ethics: The moral principles that govern our behavior. Ethics is important for the development and use of AI systems.

Explainability: The ability to explain how an AI system makes decisions. Explainability is important for building trust in AI systems.

Generative Artificial Intelligence (GAN): A type of AI that can create new content, such as images and text. GANs are used in many AI applications, such as art generation and machine translation.

Humanoid: A robot that is designed to resemble a human being. Humanoid robots are used in a variety of applications, such as manufacturing and healthcare.

Internet of Things (IoT): A network of physical devices that are connected to the internet and can collect and exchange data. The IoT is used in a wide range of applications, such as smart homes and industrial automation.

Machine Learning (ML): A field of AI that allows computers to learn without being explicitly programmed. ML algorithms learn from data and are able to improve their performance over time.

Natural Language Processing (NLP): A field of AI that deals with the interaction between computers and human (natural) languages. NLP algorithms are used in many AI applications, such as machine translation and speech recognition.

Neural Networks: A type of AI algorithm that is inspired by the structure and function of the human brain. Neural networks are used in many AI applications, such as image recognition and natural language processing.

Neurons: The basic unit of a neural network. Neurons are connected to each other and process information in a similar way to human neurons.

Prompt: A piece of text that is used to guide the output of a generative AI system. Prompts can be used to generate new text, images, and other types of content.

Prompt Engineering: The process of designing prompts for generative AI systems. Prompt engineering is important for generating high-quality and relevant content.

Reinforcement Learning (RL): A type of machine learning in which an AI agent learns to behave in an environment by trial and error. The agent receives rewards for taking actions that lead to desired outcomes and penalties for taking actions that lead to undesired outcomes. Over time, the agent learns to take actions that maximize its rewards.

Robotics: The field of engineering that deals with the design, construction, operation, and application of robots. Robots are machines that are programmed to perform tasks automatically.

Self-Aware Artificial Intelligence (SAI): A hypothetical type of AI that would be aware of its own existence and would be able to think and reason like a human being. SAI is a topic of much debate and speculation, and it is not clear whether it will ever be possible to create SAI.

Singularity: A hypothetical point in time at which technological growth becomes uncontrollable and irreversible, resulting in unforeseeable changes to human civilization. The singularity is often associated with the development of ASI, but it is possible that other technological advances could also lead to the singularity.

Supervised Learning: A type of machine learning in which the training data includes both the input and the desired output. The AI system learns from this data to predict the output for new inputs.

Theory of Mind Artificial Intelligence: The ability to understand that other people have their own thoughts, feelings, and beliefs and that these may be different from our own. Theory of mind AI is a perceived AI ability to interact with humans and other AI in a social context.

Training Data: The data that is used to train an AI system. Training data can be labeled or unlabeled. Labeled training data includes both the input and the desired output, while unlabeled training data only includes the input.

Unsupervised Learning: A type of machine learning in which the training data only includes the input. The AI system learns from this data to find patterns and relationships in the data.

Virtual Reality (VR): A simulated experience that can be similar to or completely different from the real world. VR is created using computer software and is typically experienced through a headset or goggles.

Weights: In neural networks, weights are numerical values that are assigned to the connections between neurons. The weights determine how much influence one neuron has on another. Weights are adjusted during training to improve the performance of the neural network.

References

Aggarwal, A. (2018, February 13). *The birth of AI and the first AI hype cycle*. KDnuggets. https://www.kdnuggets.com/2018/02/birth-ai-first-hype-cycle.html

AI Uncovered. (2023a, June 21). *How powerful will AI be in 2030?* [Video] YouTube. https://www.youtube.com/watch?v=XKMvk5hWDfo

AI Uncovered. (2023b, July 13). *The 6 stages of AI (beyond singularity)*. [Video] YouTube. https://www.youtube.com/watch?v=I0P2ZREqopo

AI Uncovered. (2023c, July 21). *The 5 types of AI* [Video]. YouTube. https://www.youtube.com/watch?v=oBN8EYiQleY

AI Uncovered. (2023d, July 23). *10 things they're not telling you about the new AI* [Video]. YouTube. https://www.youtube.com/watch?v=qxbpTyeDZp0

AI Uncovered. (2023e, September 3). *AI is changing education forever* [Video]. YouTube. https://www.youtube.com/watch?v=M6nPmytC99Y

AI Uncovered. (2023f). *The 7 stages of AI* [Video]. YouTube. https://www.youtube.com/watch?v=PG8vu0i5juY

Ajao, E. (2023, March 30). *The call for an AI pause points to a major concern*. Enterprise AI; TechTarget. https://www.techtarget.com/searchenterpriseai/news/365534127/The-call-for-an-AI-pause-points-to-a-major-concern

Al Jazeera English. (2023, August 19). *Robot Sophia: "Not a thing" could stop a robot takeover | Talk to Al Jazeera* [Video]. YouTube. https://www.youtube.com/watch?v=bJjXq6Pj0_c

Anderson, J. (2022). *AI, man & God | Prof. John Lennox* [Video]. YouTube. https://www.youtube.com/watch?v=17bzIWIGH3g

Arkenberg, C. (2023, June 29). *Generative AI is already disrupting media and entertainment*. Deloitte Insights. https://www2.deloitte.com/us/en/insights/industry/technology/generative-ai-tools-media-entertainment.html

AV Content Team. (2023, April 21). *This is how experts predict the future of AI.* Analytics Vidhya. https://www.analyticsvidhya.com/blog/2023/04/future-of-ai/

Bachinskiy, A. (2019, February 21). *The growing impact of AI in financial services: Six examples.* Medium; Towards Data Science. https://towardsdatascience.com/the-growing-impact-of-ai-in-financial-services-six-examples-da386c0301b2

BBC News. (2023, April 21). *AI: What is the future of artificial intelligence? - BBC News* [Video]. YouTube. https://www.youtube.com/watch?v=LK5j3pp0Too

Bennett, M. (2023, May 25). *The future of AI: What to expect in the next 5 years.* Enterprise AI. https://www.techtarget.com/searchenterpriseai/tip/The-future-of-AI-What-to-expect-in-the-next-5-years

Best, M., & Rao, A. (2022, January 18). *Understanding algorithmic bias and how to build trust in AI.* PwC. https://www.pwc.com/us/en/tech-effect/ai-analytics/algorithmic-bias-and-trust-in-ai.html

Betz, S., & Whitfield, B. (2023, March 23). *7 types of artificial intelligence.* Built In. https://builtin.com/artificial-intelligence/types-of-artificial-intelligence

Biswal, A. (2023, August 21). *Top 18 artificial intelligence applications in 2023.* Simplilearn. https://www.simplilearn.com/tutorials/artificial-intelligence-tutorial/artificial-intelligence-applications

BootstrapLabs. (2017). *A brief history of artificial intelligence* [Video]. YouTube. https://www.youtube.com/watch?v=056v4OxKwll

Buchholz, K. (2023, January 24). *Infographic: ChatGPT sprints to one million users.* Statista Infographics. https://www.statista.com/chart/29174/time-to-one-million-users/

Burns, E. (2022). *What is artificial intelligence (AI)?* TechTarget. https://www.techtarget.com/searchenterpriseai/definition/AI-Artificial-Intelligence

Bycloud. (2022). *A brief history of the entire AI field, I guess* [Video]. YouTube. https://www.youtube.com/watch?v=b9chqJ2TgzA

Carlie. (2023, August 30). *AI in education: The advantages and disadvantages.* Instrucko. https://www.instrucko.com/blogs/the-advantages-and-disadvantages-of-ai-in-education

Center for Humane Technology. (2023, April 5). *The A.I. dilemma - March 9, 2023* [Video]. YouTube. https://www.youtube.com/watch?v=xoVJKj8lcNQ

Coogan, J. (2022). The epic history of artificial intelligence [Video]. In *YouTube*. https://www.youtube.com/watch?v=jekwHBl1ySU

Crouch, G. (2023, April 20). *What's behind the A.I. hype really?* Medium. https://gilescrouch.medium.com/whats-behind-the-a-i-hype-really-e5a72a97c318

Daffer, B. (2023, May 25). *Ethics, trust, and explainability in artificial intelligence (AI).* MoreThanDigital. https://morethandigital.info/en/ethics-trust-and-explainability-in-artificial-intelligence-ai/

Dartmouth. (2006). *Artificial intelligence (AI) coined at Dartmouth.* https://home.dartmouth.edu/about/artificial-intelligence-ai-coined-dartmouth

Dawkins, J. O. (2023, January 10). *AI, cannabis, and health food are among the 9 industries with promise for business owners this year.* Business Insider. https://www.businessinsider.com/fastest-growing-industries-for-business-ai-ecommerce-supply-chain-2023-1

Diamandis, P. H. (2023, June 29). *Why AI matters and how to deal with the coming change w/ Emad Mostaque | EP #52.* [Video] YouTube. https://www.youtube.com/watch?v=ciX_iFGyS0M

Digital Adoption Team. (2023, February 9). *NLP, OCR, image recognition, and other key AI definitions.* Digital Adoption. https://www.digital-adoption.com/ocr-image/

Digital Engine. (2023, September 11). *This intense AI anger is exactly what experts warned of, with Elon Musk.* [Video] YouTube. https://www.youtube.com/watch?v=b2bdGEqPmCl

Dr. Alex Young. (2023, April 16). *The top 10 best AI studying tools 2023* [Video]. YouTube. https://www.youtube.com/watch?v=DfEDCwWPl30

Duggal, N. (2023a, August 11). *Top 10 artificial intelligence technologies in 2022.* Simplilearn. https://www.simplilearn.com/top-artificial-intelligence-technologies-article

Duggal, N. (2023b, August 21). *Future of artificial intelligence: What tomorrow might look like.* Simplilearn. https://www.simplilearn.com/future-of-artificial-intelligence-article

Editorial Team. (2023, August 17). *Explained: What the hell is deep learning ?* Techovedas. https://techovedas.com/explained-what-the-hell-is-deep-learning/

Expert Panel®. (2021, June 14). *Council post: 14 ways AI could become a detriment to society.* Forbes. https://www.forbes.com/sites/forbestechcouncil/2021/06/14/14-ways-ai-could-become-a-detriment-to-society/?sh=21bfdc2b27fe

Fallon, J. (2017). *Tonight Showbotics: Jimmy meets Sophia the human-like robot* [Video]. YouTube. https://www.youtube.com/watch?v=Bg_tJvCA8zw

Gadzhi, I. (2023, August 15). *These 7 AI tools will make you rich* [Video]. YouTube. https://www.youtube.com/watch?v=-qReeg7imGc

Gillis, A. S. (2023, June). *What is responsible AI? Guidance and best practices.* SearchEnterpriseAI. https://www.techtarget.com/searchenterpriseai/definition/responsible-AI

Global Data Thematic Intelligence. (2023, April 18). *Generative AI: The reason behind the hype.* Verdict. https://www.verdict.co.uk/generative-ai-vested-interests/

Glover, E. (2023, February 7). *5 AI trends to watch in 2023.* Built In. https://builtin.com/artificial-intelligence/ai-trends-2023

Grand View Research. (2022). *Artificial intelligence market size, share | AI industry report, 2025.* https://www.grandviewresearch.com/industry-analysis/artificial-intelligence-ai-market

Griffiths, E. (2023, February 13). *Stars who had to retire from acting careers due to illness.* HELLO! https://www.hellomagazine.com/healthandbeauty/health-and-fitness/20230213164244/stars-retired-from-acting-illness/

Harari , Y. N. (2023, May 14). *AI and the future of humanity | Yuval Noah Harari at the Frontiers Forum* [Video]. YouTube. https://www.youtube.com/watch?v=LWiM-LuRe6w

Hitachi Systems Security. (n.d.). *Can AI, ethics, and privacy work in harmony?* Hitachi. https://hitachi-systems-security.com/can-ai-ethics-and-privacy-work-in-harmony/

Hiter, S. (2023, July 12). *Top 6 AI trends 2023.* EWEEK. https://www.eweek.com/artificial-intelligence/ai-trends/

Howarth, J. (2023, August 14). *57+ amazing artificial intelligence statistics (2023).* Exploding Topics. https://explodingtopics.com/blog/ai-statistics

IBM Data , & AI Team. (2023, July 6). *AI vs. Machine Learning vs. Deep Learning vs. Neural Networks: What's the difference?* IBM Blog. https://www.ibm.com/blog/ai-vs-machine-learning-vs-deep-learning-vs-neural-networks/

IBM Technology. (2021a). *What is AI ethics?* [Video]. YouTube. https://www.youtube.com/watch?v=aGwYtUzMQUk

IBM Technology. (2021b, August 11). *What is NLP (natural language processing)?* [Video] YouTube. https://www.youtube.com/watch?v=fLvJ8VdHLA0

IBM Technology. (2023a, April 10). *AI vs Machine Learning* [Video]. YouTube. https://www.youtube.com/watch?v=4RixMPF4xis

IBM Technology. (2023b, August 4). *How to add AI to your apps faster with embedded AI* [Video]. YouTube. https://www.youtube.com/watch?v=OThahaOga20

Karijan, R. (2023, August 16). *The history of artificial intelligence: Complete AI timeline.* Enterprise AI. https://www.techtarget.com/searchenterpriseai/tip/The-history-of-artificial-intelligence-Complete-AI-timeline

Kerry, C. (2020, February 10). *Protecting privacy in an AI-driven world.* Brookings. https://www.brookings.edu/articles/protecting-privacy-in-an-ai-driven-world/

Kumar, S. (2019, November 25). *Advantages and disadvantages of artificial intelligence.* Medium; Towards Data Science. https://towardsdatascience.com/advantages-and-disadvantages-of-artificial-intelligence-182a5ef6588c

Kurkina, I. (2023, August 17). *12 best AI project management software tools.* Academy SMART. https://academysmart.com/12-best-ai-project-management-software-tools/

Latinne, P. (2023, June 13). *How AI can drive productivity and value in the financial sector.* Ernst & Young (EY). https://www.ey.com/en_be/digital-transformation/how-ai-can-drive-productivity-and-value-in-the-financial-sector

Lutkevich, B. (2023, June 27). *Artificial intelligence glossary: 60+ terms to know.* WhatIs.com? https://www.techtarget.com/whatis/feature/Artificial-intelligence-glossary-60-terms-to-know

Manyika, J., Silberg, J., & Presten, B. (2019, October 25). *What do we do about the biases in AI?* Harvard Business Review. https://hbr.org/2019/10/what-do-we-do-about-the-biases-in-ai

Marr, B. (2023, April 12). *The 5 biggest artificial intelligence (AI) trends in 2023* [Video]. YouTube. https://www.youtube.com/watch?v=grmudb9FQpI

Matleena, S. (2023, August 31). *28 AI statistics and trends in 2023*. Hostinger Tutorials. https://www.hostinger.com/tutorials/ai-statistics

Matson, M. (2023, July 14). *Llm Data: Tech terms explained | blog*. Player Zero. https://www.playerzero.ai/advanced/tech-terms-explained/llm-data-tech-terms-explained

Mauran, C. (2023, April 6). *Whoops, Samsung workers accidentally leaked trade secrets via ChatGPT*. Mashable. https://mashable.com/article/samsung-chatgpt-leak-details

McKinsey. (n.d.). *Workplace automation and future jobs*. McKinsey & Company. https://www.mckinsey.com/featured-insights/artificial-intelligence/five-fifty-fear-fear-not

Melanoma Research Alliance. (2022, November 15). *Artificial intelligence & melanoma detection: Closing the gaps*. https://www.curemelanoma.org/blog/article/artificial-intelligence-and-melanoma-detection-closing-the-gaps

Musienko, Y. (2023, June 1). *Biggest artificial intelligence (AI) trends in 2023*. Merehead. https://merehead.com/blog/artificial-intelligence-2023/

Mylett, E. (2023, August 8). *Is artificial intelligence our "Oppenheimer Moment"? Mo Gawdat's warning to the world* [Video]. YouTube. https://www.youtube.com/watch?v=Ae4o92F4tVM

NASA. (2019). *NASA's Spirit and Opportunity Mars Rovers*. https://mars.nasa.gov/mer/

NASA Jet Propulsion Laboratory. (2022, March 15). *NASA's Mars Rovers Are On the Move and Bringing the Public Along (NASA Mars Report March 15, 2022)* [Video]. YouTube. https://www.youtube.com/watch?v=8wiOJsKdz04

Nazer, L., Zatarah, R., Waldrip, S., Xue Chen Ke, J., Moukheiber, M., Khanna, A. K., Hicklen, R. S., Moukheiber, L., Moukheiber, D., Ma, H., & Mathur, P. (2023). Bias in artificial intelligence algorithms and recommendations for mitigation. *National Library of Medicine, 2*(6), e0000278–e0000278. https://doi.org/10.1371/journal.pdig.0000278

Pattam, A. (2021, June 7). *7 ethical issues with AI that you should know about* [Video]. YouTube. https://www.youtube.com/watch?v=KyB7NSWEODE

Payne, L. (2023, September 5). *Deepfake | history & facts*. Britannica. https://www.britannica.com/technology/deepfake

Perri, L. (2023, August 17). *What's new in artificial intelligence from the 2023 Gartner Hype Cycle*TM. Gartner. https://www.gartner.com/en/articles/what-s-new-in-artificial-intelligence-from-the-2023-gartner-hype-cycle

Petersson, D. (2023, June 21). *AI vs. machine learning vs. deep learning: Key differences*. Enterprise AI. https://www.techtarget.com/searchenterpriseai/tip/AI-vs-machine-learning-vs-deep-learning-Key-differences

Pramoditha, R. (2022, March 21). *The relationship between AI, ML, NNs and DL*. Data Science 365. https://medium.com/data-science-365/the-relationship-between-ai-ml-nns-and-dl-60bd40069908

Rahulkwh. (2023, April 18). *10 best AI tools to boost productivity in 2023*. GeeksforGeeks. https://www.geeksforgeeks.org/10-best-ai-tools-to-boost-productivity/

RisingMax. (2023, September 4). *How much does it cost to build artificial intelligence in 2023*. https://risingmax.com/blog/how-much-does-artificial-intelligence-cost

Robbins, T. (2020). *Meet Sophia, world's first AI humanoid robot | Tony Robbins* [Video]. YouTube. https://www.youtube.com/watch?v=Sq36J9pNaEo

Sajid, H. (n.d.). *AI in robotics: 6 groundbreaking applications*. V7labs. https://www.v7labs.com/blog/ai-in-robotics

Sajid, H. (2023, April 1). *AI in cybersecurity: 5 crucial applications*. V7labs. https://www.v7labs.com/blog/ai-in-cybersecurity

Sakshiparikh23. (2023, June 5). *Difference between machine learning and deep learning*. GeeksforGeeks. https://www.geeksforgeeks.org/difference-between-machine-learning-and-deep-learning/

Satariano, A., & Metz, C. (2023, March 5). Using A.I. to detect breast cancer that doctors miss. *The New York Times*. https://www.nytimes.com/2023/03/05/technology/artificial-intelligence-breast-cancer-detection.html

Simplilearn. (2023, February 23). *How does artificial intelligence (AI) work and its applications*. https://www.simplilearn.com/tutorials/artificial-intelligence-tutorial/how-does-ai-work

Somers, M. (2020, July 21). *Deepfakes, explained.* MIT Sloan. https://mitsloan.mit.edu/ideas-made-to-matter/deepfakes-explained

Spisak, B., Rosenberg, L. B., & Beilby, M. (2023, June 30). *13 principles for using AI responsibly.* Harvard Business Review. https://hbr.org/2023/06/13-principles-for-using-ai-responsibly

Tableau. (n.d.). *What are the advantages and disadvantages of artificial intelligence (AI)?* https://www.tableau.com/data-insights/ai/advantages-disadvantages

Tantsiura, P. (2023, August 28). *Best use cases of AI, machine learning in the transportation industry.* The App Solutions. https://theappsolutions.com/blog/development/ai-in-transportation/

The Economic Times. (2023, April 25). *AI and privacy: The privacy concerns surrounding AI, its potential impact on personal data.* https://economictimes.indiatimes.com/news/how-to/ai-and-privacy-the-privacy-concerns-surrounding-ai-its-potential-impact-on-personal-data/articleshow/99738234.cms?from=mdr

This Morning. (2023). *Meet Ameca! The world's most advanced robot* [Video]. YouTube. https://www.youtube.com/watch?v=vE9tIYGyRE8

Thomas, M. (2021, July 21). *The future of artificial intelligence.* Built In. https://builtin.com/artificial-intelligence/artificial-intelligence-future

Thomson Reuters. (2023). *Addressing bias in artificial intelligence the current regulatory landscape.* https://www.thomsonreuters.com/en-us/posts/wp-content/uploads/sites/20/2023/08/Addressing-Bias-in-AI-Report.pdf

UN News. (2023, July 10). *Meet the robots who are making the world a better place.* Africa Renewal. https://www.un.org/africarenewal/magazine/july-2023/meet-robots-who-are-making-world-better-place

Venture City. (2021, March 6). *The first 10,000 days on Mars (timelapse)* [Video]. YouTube. https://www.youtube.com/watch?v=G3hPH_bc0Ww

Wall Street Journal. (2021, December 21). *What NASA's Perseverance Rover has learned after 10 months on Mars | WSJ* [Video]. YouTube. https://www.youtube.com/watch?v=WrTHX8t0yl8

WatchMedia. (2020). *Robots evolution | 100BC - 2020* [Video]. YouTube. https://www.youtube.com/watch?v=Uzf-K7UtD78

Wheeler, T. (2023, June 15). *The three challenges of AI regulation*. Brookings. https://www.brookings.edu/articles/the-three-challenges-of-ai-regulation/

Yuanyuan, F. (2021, July 1). *What is Embedded AI (EAI)? Why do we need EAI?* Huawei. https://info.support.huawei.com/info-finder/encyclopedia/en/EAI.html

Zauderer, S. (2023, July 23). *Artificial intelligence statistics & facts for 2023*. Cross River Therapy. https://www.crossrivertherapy.com/research/artificial-intelligence-statistics

Milton Keynes UK
Ingram Content Group UK Ltd.
UKHW011257221123
433051UK00008B/477

9 798989 429004